Presenting Service

THE ULTIMATE GUIDE FOR THE FOODSERVICE PROFESSIONAL

Second Edition

Lendal H. Kotschevar, Ph.D., FMP

Valentino Luciani, CHE

John Wiley & Sons, Inc.

This book is printed on acid-free paper. ♾

Published by John Wiley & Sons, Inc., Hoboken, New Jersey
Published simultaneously in Canada

For general information on our other products and services or for technical support, please contact our Customer Care Department within the United States at (800) 762-2974, outside the United States at (317) 572-3993 or fax (317) 572-4002.

Wiley also publishes its books in a variety of electronic formats. Some content that appears in print may not be available in electronic books. For more information about Wiley products, visit our web site at www.wiley.com.

Library of Congress Cataloging-in-Publication Data:
Kotschevar, Lendal Henry, 1908–
Presenting service : the ultimate guide for the foodservice professional / Lendal H. Kotschevar, Valentino Luciana.—2nd ed.
p. cm.
Includes bibliographical references and index.
ISBN-13: 978-0-471-47578-1 (pbk.)
ISBN-10: 0-471-47578-5 (pbk.)
1. Food service. 2. Food service employees—Training of. I. Luciani, Valentino. II. Title.
TX911.K66 2006
642′.6—dc22 2005056823

Printed in the United States of America

16 15 14

This book is dedicated to
my daughter, Julie Kotschevar,
who helpfully made this revision possible

CONTENTS

Service Mise en Place

Service in Various Industry Segments

Service Areas and Equipment

Classic Service Styles

Serving the Meal

Bar and Beverage Service

PREFACE

Every author, in writing a book, has a purpose. We had two: (1) to detail what managers and servers should know and do to serve foods and beverages competently, and (2) to indicate how to make a professional career out of serving others. Too frequently, the act of service is considered a menial job, one that people do only to make money until a better and more respected job comes along. On the contrary, the foodservice industry enjoys among its millions of employees many experienced, talented, professional servers who contribute much to the industry's overall success.

Serving people is sometimes difficult and demeaning work, but the rewards outweigh the challenges. Education, training, and a professional attitude are the ingredients needed to harvest those rewards. With this revised comprehensive textbook, we hope to educate future servers, supervisors, and managers in the techniques and demeanor of professional service.

Presenting Service: The Ultimate Guide for the Foodservice Professional, Second Edition, covers the basics, as well as advanced topics, that future foodservice managers need to know to give successful table and customer service. It covers the historical context of service, the manager's role in good service, and the various types of service in foodservice operations. Three new chapters have been added, covering bar and beverage service, table etiquette, and classic service styles.

We are confident that the information presented will help build essential skills for a successful foodservice career. Excellent customer service is crucial to the success of any operation yet can be frequently overlooked by managers and employees. This resource is an invaluable tool to help build better service practices throughout the industry.

What's New for the Second Edition

Many important changes and additions have been made to **Presenting Service: The Ultimate Guide for the Foodservice Professional, Second Edition,** to make this book even more useful.

- Extensive **case studies** and real-world scenarios are included with every chapter that enables readers to apply the concepts presented.
- A full chapter on **bar and beverage service** is new to this edition and includes specialty coffees, cocktails, and wine service, enabling the service and recommendation of beverage alcohol and nonalcoholic beverages to guests based on informed knowledge.
- **Classic service styles**, including the techniques of French, Russian, American, English, and Chinese service, are described in a separate new chapter.
- A new chapter on **table etiquette** contains a historical perspective as well as a complete description of etiquette rules concerning special foods and various cultures, such as European, Chinese, Indian, and Middle Eastern dining.
- **Customer service and foodservice security** boxes convey tips and best practices for handling customer and security issues throughout the book.
- This book contains **checklists** throughout that easily walk the server and manager through good service practices.
- An expanded new **glossary** includes additional key terms for bar and beverage service, table etiquette, and classic service styles.
- A new appendix, **Duties of Some Service Workers**, explains the responsibilities of the frontline staff, including the host, server, bus person, and bar server.
- **Internet sources** at the end of selected chapters provide Web sites for additional information presented in the chapter.

Just as a master craftsman takes a raw, crude stone and turns it into a beautiful, sparkling gem, so does an editor take a manuscript and turn it into a published book. The authors would like to thank the editors at John Wiley & Sons for their assistance in the development of this book.

LENDAL H. KOTSCHEVAR, PH.D., FMP
VALENTINO LUCIANI, CHE

ACKNOWLEDGMENTS

The input of the advisors and reviewers for this and the previous edition has been invaluable. We wish to thank the following people for their contributions:

Adam Carmer, University of Nevada, Las Vegas
Celia Curry, Art Institute of New York
Ed Debevic's Deerfield Restaurant
G. Michael Harris, Bethune-Cookman College
David Hoffman, Mohawk Valley Community College
Mike Jung, Hennipen Technical College
Jeanette Kellum, Romano Brother Beverage Company
Peter Kilgore, The National Restaurant Association
Lettuce Entertain You Enterprises, Inc.
Debra Orsi, Art Institute of Chicago
Vickie Parker, Brinker International, Chili's
Lawrence D. Posen, FMP, Eurest Dining Services
Arthur Riegal, Sullivan County Community College
Edwin Rios and the staff of the Palmer House Hilton
Vincent Rossetti, Nordstrom's, Oak Brook
Sean Ryan, Art Institute of Los Angeles
Marcia Shore, Harrisburg Area Community College
Peter Simoncelli, Four Seasons Hotel, Chicago
Will Thorton, St. Philip's College
Diane Withrow, Cape Fear Community College
Mike Zema, Elgin Community College

A Historical Overview of Service

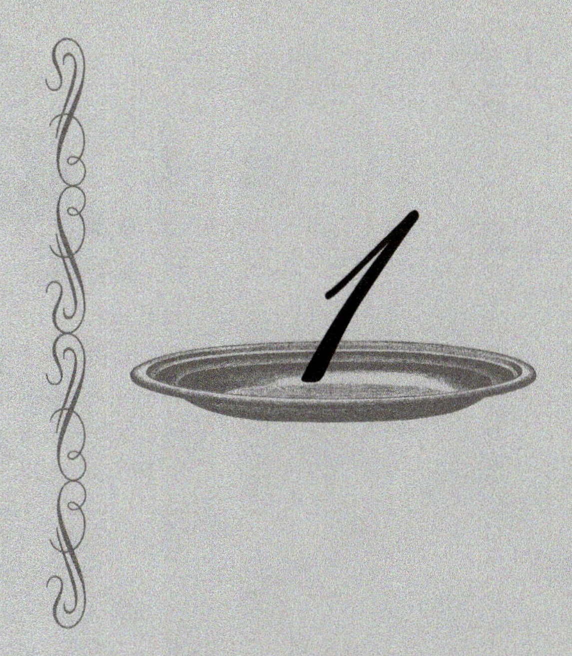

Outline

I. Introduction
II. The Age of Service
III. Service: A Total Concept
   A. The Meaning of Service
      1. The Tradition of Hospitality
      2. Meeting and Exceeding Guest Expectations
IV. A Historical Overview of Service
   A. Beginnings
      1. Ancient Times
      2. Greek and Roman Times
      3. Street Vending
      4. Service in Inns and Taverns
   B. The Dignification of Dining
      1. Well Being
      2. Religion
      3. Socializing and Recreation
   C. The Development of European Haute Cuisine
      1. The Gastronomic Influence of Catherine de Medici
      2. The *Restorante*
      3. Discriminating Gourmets
   D. The Growth of Service in Modern Times
      1. The Rise of Hotels
      2. Restaurants and Service in the United States
V. Chapter Summary
VI. Chapter Review
VII. Case Studies

Learning Objectives

After reading this chapter, you should be able to:

- Describe the importance of excellent service to a successful operation.
- Provide a historical overview of service.
- Explain how haute cuisine developed, and how it influenced modern service.

INTRODUCTION

Excellent service is vital to the success of every foodservice operation. Many operations fail not because the food or atmosphere are inadequate, but because the service fails to please guests. The National Restaurant Association has reported that 49 percent of all customer complaints involve service, compared to 12 percent for food, 11 percent for atmosphere or environment, and 28 percent for other reasons.

Fifteen years ago, American selected restaurants based on the following, by rank:

1. Quality of service
2. Quality of food
3. Ambiance
4. Price and value relationship
5. Parking and accessibility
6. Various other factors

According to a 2005 report by the National Restaurant Association, restaurant selection has been redefined. Currently trends for choosing a dining facility are:

1. Cleanliness of the premises (sanitary standards)
2. Quality of service
3. Quality of food
4. Price and value relationship

The complete subject of cleanliness is too extensive to discuss here. For those interested in an extensive treatment of sanitation refer to the National Restaurant's *Applied Food Service Sanitation.*

Excellent service depends on excellent, professional servers who not only know their jobs and perform them well, but understand their guests and how to best meet and exceed their needs. This includes the ability to work with others as a team to deliver great service, and the attitude to approach the job as a professional.

Serving is not an easy job. It requires hard work, time to learn to do it well, and a commitment to serving people.

FIGURE 1.1

Excellent service is just as important in a dining operation as the food.

THE AGE OF SERVICE

Not long ago, the economies of the world's most advanced nations were based largely on the production of goods. This is no longer true. Increased productivity, disposable income, and leisure time have contributed to a growing demand for service industries, so much so that we say we are in the *Age of Service.*

Serving food and beverages is a significant part of a huge and profitable industry in the dominant service sector. The National Restaurant Association estimates that yearly foodservice sales in the 900,000 eating and drinking places in the United States were over $437 billion 2003, nearly 4 percent of the U.S. gross domestic product. (See **Exhibit 1.1.**) The foodservice industry employs more than 12 million people, making up 9 percent of the work force. Nearly one-half of all the adults living in the United States eat out at least once a day.

EXHIBIT 1.1 Projected Sales 2005—Foodservice Industry

2005 Sales	**(Billion $)**
Commercial	437
Eating places	326
Drinking places	15
Managed services	32
Hotel/motel restaurants	25
Retail, vending, recreation, mobile	39
Other	39

SOURCE: *National Restaurant Association*

As the economy changes, people are finding that work in the service sector offers good and permanent opportunities. The foodservice industry is essential to this economy and will continue to grow, probably at a greater rate than many other service industries. Joining the service staff in the foodservice industry can provide a permanent position that pays well and gives adequate job benefits.

The foodservice industry is thriving, and highly competitive. What differentiates one foodservice establishment from another? It is often a distinctive and excellent reputation for service. Food services have found that price wars to meet competition usually do not work, but raising the level of service can be highly effective in rising above competition. People who dine out are much more service-sensitive than they used to be and will often select where they will dine by the level of service given. Food services have found that it costs very little more to provide good service rather than poor service.

Service: A Total Concept

The Meaning of Service

What do we mean by *service?* It is more than taking orders, placing down food and beverages, and clearing up after a meal. It is the act of providing customers with a wide range of meal-related benefits and experiences. Service is what makes people feel good about spending their money to eat out.

Serving should not be looked upon as menial. Too frequently, servers downgrade their work. This is because they fail to understand what their task is and do not realize that serving can be professional work.

But, is serving a profession? The answer is yes. One definition of a profession is qualified persons in one specific occupation or field. We can be more precise if we add the phrase "serving the needs of others," to then say qualified persons of one specific occupation or field serving the needs of others.

Many professional people have positions that require them to serve others. A doctor serves the sick. A religious leader serves those in need of spiritual guidance. A dietitian helps others to select healthful foods. In many cultures, teaching is a highly respected service profession. Food servers meet the needs of others by serving their needs. Thus, those who serve food and drink are professionals in that they are a large body of qualified people working in one occupation, serving others' needs.

Being a professional brings on responsibilities. Professional people are supposed to meet the highest standards of moral and ethical behavior. They are expected to treat others in a professional manner. Those who serve should be proud of their work. Servers who approach their jobs professionally and are proud to serve others enhance the industry as well as their own careers. Mastering the art of service builds pride and self-esteem, and opens up a world of career opportunities.

The Tradition of Hospitality

Hospitality encompasses two important concepts: Guests should always be made to feel welcome and wanted, and all efforts should be made to see that no guest comes to any harm. These are ancient rules of custom in nearly every culture.

Many ancient peoples formalized ways in which guests were to be received when they came to one's home. An old Irish custom was to offer a pinch of salt and a small glass of wine when guests came to visit, both wine and salt being precious commodities. In ancient times, Jewish people greeted their guests by bathing their feet and rubbing them with fragrant oil. The Chinese offered special foods and drink to guests.

Another social rule that developed many years ago was that when guests were in the premises they should be protected from any harm. The concept of **sanctuary** was especially important to the early Christians, whose churches, monasteries, and convents were recognized as places of protection even from government or royal authorities. This feeling of sanctuary strongly influenced the rules of how guests should be treated at inns and restaurants. This old European value has evolved into modern laws holding innkeepers especially liable for the safety of guests. The concept has been extended, to restaurants and other hospitality operations.

Meeting and Exceeding Guest Expectations

Service is often the single most influential factor in customers' decisions as to where to eat out. Great service gives operations a competitive edge, and keeps people coming back. A good server must learn how to read each guest to determine how to meet particular needs, and how to exceed guests' expectations.

Good servers do three things well: They pay close attention to detail, they work efficiently, and they are consistent even when a dozen things go wrong and threaten their demeanor. They seek the rewards—good tips, higher wages, recognition from their peers and employers—of focusing completely on the details of their work. To be efficient, servers need not kill themselves with hard work, or be rude or abrupt with customers. Instead, they must learn to plan and organize to make the best use of their time by doing the following:

- Set up work stations carefully at the start of a shift so all supplies are available.
- Replace supplies before they run out.
- Don't walk from one area to another empty-handed if there is something to carry.
- Combine trips.
- Stay organized.
- Follow the most efficient routine.
- Save steps whenever possible.
- Prepare for busy times.
- Stay on top of the job during slower times.

Today's guests are quite sophisticated. They expect good service, so the challenge is to impress them by exceeding their expectations. To do this, servers must:

- Focus completely on customers.
- Show a sense of urgency.
- Acknowledge, greet, and say goodbye to every customer with whom they come in contact.

Good servers also must anticipate guests' needs, and try to accommodate them before they think to ask. This means watching and listening to customers carefully for clues as to what their needs might be, doing whatever is reasonably possible to please them, and thinking creatively when serving customers. For example:

- If a customer is standing at a quick-service operation's counter staring at the menu, a server should suggest several items, or ask if the customer has any particular questions about the menu.
- If guests in a full-service operation slow down, pause, and look around the dining room as they and their host(ess) approach a table, the host(ess) should ask if that table will be all right.
- If customers come into a quick-service restaurant with a small child but do not order food for the child, the server should ask whether they want an extra set of utensils or any appropriate children's items (coloring place mats, etc.).
- Any time a customer is looking around confusedly, a server should ask whether they need help finding something.
- A server whose customers are writing on a napkin should ask if they would like some paper.

A Historical Overview of Service

The growth of service in food establishments is not well documented, especially in its early stages. What it was and how it grew must be gleaned from brief references in literature. In *The Canterbury Tales,* Geoffrey Chaucer writes of a nun wiping her lips daintily with a napkin, from which we can infer that fourteenth-century Englishwomen used napkins. Another source of how service developed is to note events or practices among people of the times. Thus, from evidence about the nearly one hundred different dishes served at a formal seventeenth-century dinner in France, and the elegant tableware used, we infer that elaborate service must have marked these royal events.

Beginnings

Ancient Times

Early people ate largely for survival. There was little ceremony involved. With the discovery of fire, some foods were cooked. Clay was used to make dishware and other utensils that could hold food while it was cooked over the fire. Thus, the diet changed from raw foods to stewed and roasted meat, cooked seeds, vegetables, and other items. Many of these ancient cooking pieces have been discovered. We find that the earliest pieces are crudely made, but gradual improvements were made in the clay mixtures used, and their design. People enlarged and perfected the kinds of ware used and began to make pieces from which to eat and drink. They found out how to color and glaze this ware. In some cases, ladles and cooking spoons were made. These improvements were undoubtedly a boost to their level of service.

After humans moved from caves and built dwellings, fireplace cooking developed. This was an advancement in cooking technique but service remained crude and rudimentary. Excavations in the Orkney Islands near Denmark show that around 10000 B.C., people built their dwellings around a common kitchen and cooked their food and ate together as a communal group. There is no evidence of eating utensils.

Diggings from somewhat later times in the Mohenjo-Daro region in modern Pakistan reveal the existence of restaurant-type units where the public went to dine. The ancient Chinese also had restaurants that served food and drink in fine pottery and porcelain dishware. It is thought that the Chinese have used chopsticks since 6000 B.C. It was not until six or seven thousand years later that the knife, fork, and spoon, as a place setting, were developed somewhere in southwest Asia.

Greek and Roman Times

The Romans had small eating and drinking establishments called ***taberna vinaria,*** from which we get the word *tavern.* These *tabernas* were so popular in the third century B.C. that they were found tucked away into every corner of every large city. Ruins in the volcanically preserved city of Pompeii tell us that diners could eat and drink in restaurants that featured stone counters outside and stone benches and tables inside. Cooked food was kept warm in ***thermopoliums,*** stone counters with holes in them to keep food warm. The *tabernas* dispensed a significant amount of wine. Huge stone jars contained the wine, which was preserved by pouring oil over the top to prevent air from contacting it and turning it into vinegar.

There are many accounts of pagan feasts in honor of the gods of ancient times. From the vast amount of food and drink prepared and the numbers that attended, a fairly high level of catering service must have been developed. Greeks held feasts and celebrations that lasted for several days. Their feast to Dionysus, the god of wine, was one of great rejoicing, revelry, and excitement.

The ancient Romans rivaled the Greeks in their use of elaborate public feasts. One Roman emperor bankrupted the state treasury by stealing money for his private feasts. The wealthy and influential also gave sumptuous private feasts for large numbers of their friends, spending huge sums on them. The service was elaborate. A typical banquet had four courses. The first was called *mensa prima,* the second *mensa secunda.* Many different kinds of rare and exotic foods from all over the Roman Empire were elegantly served. Tableware included beautiful glassware and ceramic and metal dishware. Hosts vied with each other to see who could put on the most elaborate banquets. One Roman, named Apicius, spent so much money on one that he bankrupted himself and committed suicide. He wrote the first cookbook that we know of, and today his recipes are still used in many food services.

Street Vending

Street vending service also developed early. Wall paintings in ancient Egyptian tombs show vendors selling food in markets, where people ate standing in the street. Vending and street eating were also common in ancient China, when a vendor would come down a lonely, dark street at night crying out his menu. People would come out of their homes and make a purchase. Vending of this kind is still done in many countries.

Service in Inns and Taverns

Inns and taverns were established in many ancient countries to care for land and sea travelers. China and India had laws regulating inns. Another law in China required monasteries to provide care for travelers. The writings of the thirteenth-century Italian explorer Marco Polo describe his stays at Chinese inns and monasteries. The ruins of some of these inns are still to be found in the dry desert areas along the ancient Silk Route. Wherever enough travel developed along these ancient routes, a hostel or inn was certain to be found.

For the most part, these units offered limited service. Beds and space for travelers' animals was provided; some offered food and libations. Usually, servers served the food, but in some inns, food was prepared by travelers or travelers' servants.

The Dignification of Dining

Ancient peoples also did much to dignify and formalize food service. Instead of just making eating an exercise in gaining sustenance, they began to attach to it philosophic or symbolic meanings, so that in the act of eating they were expressing a feeling or belief. These practices greatly influenced the kind of food served, how it was prepared, and how it was served. These customs arose around the concepts of well being, religion, entertainment, and social reasons. Very few cultures failed in one way or another to develop such practices in these areas.

Well Being

A number of cultures selected certain foods to eat primarily for health or sanitary reasons. Some beliefs were so firmly held that even though the results were not positive, the practices were continued with religious zeal.

However, many food remedies were effective. Today, we know much more about the need for a nutritional diet. We have the benefit of scientific knowledge, but in ancient times people learned by trial and error what one should serve, and it is surprising how well some cultures did in achieving a diet that led to better health.

The Chinese led the way in establishing rules for achieving good health through food. To the Chinese, food was medicine, and medicine was food, since both nourished the body. Confucius established strict rules as to what foods to serve, how to combine foods, the amount to serve, and when foods should be eaten. More than 2,800 years ago, the Chinese emperor Shennung wrote a cookbook, the *Hon-Zu,* that is still used. The art of eating well has been part of many Asian cultures for millennia.

The Chinese, and many other cultures, believed that when one ate certain foods, one took on the characteristics of the source of the food; eating tiger meat could make one fierce and aggressive, or eating an eye or a liver would make these organs in the body stronger. Even today, cultures of the world believe that the service of certain foods can bring about desirable results. Today, for example, it is the custom of many southern people in this country to eat black-eyed peas on New Year's Day to ensure good luck for the rest of the year.

A number of cultures also used the service of food and drink to signify delicate social relationships such as respect, love, contempt, devotion, or other feelings. The service of cold noodles to a guest in China indicates a lack of warmth for the relationship but, if served warm, indicates deep respect. If a Hopi Indian woman wanted to indicate a romantic interest in a man, she would give him two small pieces from a maiden's cake, made of blue corn meal and filled with boiled meat. To show matrimonial interest, she placed this on a plate of blue corn flat bread outside the door of the man's house. If he too had an interest, he took the plate inside, but if not, the plate was left outside and some family member of the girl retrieved it so she would not be embarrassed by having to remove it herself.

Religion

The Jewish people probably developed the most complete set of religious practices using the service of food to symbolize the practices. In fact, one section of the Torah is given over completely to a delineation of dietary, or **kosher,** laws. Certain rules are followed in kosher dietary laws. Only mammals that have split hooves and chew their cud are allowed. This excludes pork. Only a specific list of birds, including most birds commonly eaten, are allowed. Both mammals and birds must be slaughtered in a specific manner by a *shochet,* a trained Jewish slaughterer. Meat and poultry must also be *koshered,* or soaked and salted to remove all blood, which is forbidden. Only fish with fins and scales are allowed. This excludes all shellfish. Cooking on the Sabbath is forbidden, so food must be prepared in advance and eaten cold or heated without direct contact to fire. Milk and meat cannot be eaten together or even at the same meal. Kosher rules completely regulate dining.

Even today, people of the Jewish faith practice customs that include dining restrictions. For example the *Seder* dinner on the first two evenings of Passover and the eating of matzo or unleavened bread during Passover symbolize the escape of the Jewish tribes from Egypt, a time in which they had no chance to prepare the leavened product. The serving of *haroset,* a mixture of nuts, fruits, wine, and spices, symbolizes the mortar the enslaved Jews were forced to use to build the Pharaoh's pyramids. Fresh parsley or other vegetables call to mind their hopes of spring and freedom. Bitter herbs are served to symbolize the years as slaves in Egypt. Saltwater represents the people's tears during that time.

Similarly, The Koran describes Muslim dietary laws, allowing only food that is **halal,** which is Arabic for permitted or lawful. Those who observe *halal* do not consume pork, carniverous animals, birds of prey, land animals without external ears, most reptiles and insects, blood, or alcohol and intoxicants. Permitted animals must be slaughtered in a specific manner. Ramadan, the ninth month of the Islamic year, is still spent in fasting from sunrise to sunset. There are a number of rules that establish customs, such as the washing of hands before dining and the proper use of the hands in eating.

The Buddhist religion also established a number of practices. Eating meat at certain times is forbidden, making room for a wide number of different foods with special service requirements. On feast days, food is brought to the temples and laid at the feet of the statue of Buddha. Hindu mythology relates how Prajapati, the Lord of Creation, created *ghee,* or clarified butter, by rubbing butter in his hands over a fire, dropping some into the fire. He discovered the heat of the fire drove off the liquid in the butter. The people of India still consider ghee precious and ritualistically reenact Prajapati's act of creation by pouring it over a fire.

The rice farmers of Bali still practice ancient customs. The growing of rice, their basic food, has been woven into their religion. A group of farmers using water from a dam join together in a group called a *tempek.* This group worships and works together in the fields. They have two temples: one in the fields, and one near the dam. These temples contain their ritual calendars that indicate the time of planting, harvesting, and other activities. They have a main temple on the island's only mountain and two more near a lake in the island's center. Delegates from all tempeks meet every 210 days at the main temple to perform rituals, celebrate, and mark the intertwining of their lives with the cultivation of their rice.

The Christian religion uses food and drink to symbolize theological tenets. The use of bread and wine to symbolize the body and blood of Christ is an example. Easter and Christmas are celebrated by the service of special foods and drinks.

Socializing and Recreation

Food and beverages were used to support public entertainment and social affairs. The Greeks often used their eating places as a sort of club where they could gather and talk together about common affairs while they ate and drank. The word *colloquium* comes from the Greek word meaning to gather together to eat and drink.

Many other early cultures also used food and beverages to promote social life and entertainment. China had wine shops where people gathered to drink. Several of their greatest poets wrote their poetry there, reading it out loud to other guests. In Europe the tavern acted as a similar gathering place where people could meet socially. In the Arab countries people gathered together to dine and be entertained by acrobatic feats.

Back then, there were no electric lights, no radios, and no televisions. Newspapers, magazines, and other printed matter had not been introduced. Transportation was limited, and few ever left the area where they were born. Food was their major concern, and it was natural that they would use it as a way to enrich and extend their lives.

In ancient Egypt, meals were often simple, yet important occasions at which family, friends, neighbors, and even traveling strangers were welcome. Egyptian people ate bread, cured fish from the Nile and its tributaries, and cooked leeks and onions with meat, small game, and birds. Usually the meal was accompanied by barley wine or sweet fruit wine. Though the foods of various classes were similar, their tableware differed. Poor people ate out of glazed pottery and dishes, while the rich ate from metal dishes, used ivory and wooden spoons, and drank from glass goblets. The poor usually drank barley wine, while the rich drank fruit wine. Unlike Greeks and Romans, who often dined according to gender (a custom that continued in Europe until this century and a custom that still exists today in some south Asian nations), Egyptian women and men dined together.

In almost every ancient culture of the world, food and beverages were used as a means of worship or reverence. The Japanese tea ceremony has strong overtones of religious worship. The previous restriction against eating meat on Fridays in the Roman Catholic faith was viewed as an act of reverence to Jesus Christ. The refusal to eat meat by the Buddhists and others of different faiths is a further example of ritualistic eating.

In all of these customs, people would pray or otherwise indicate in their reverence their deep belief in what the service symbolized. As a result, the service of certain foods and beverages became very important by symbolizing faith and religion.

Pagan cultures also developed similar customs of reverence. Some feasts to the gods were marked by drinking wine and celebration. Often animals were sacrificed, or specially prepared and served. The use of chicken or other fowl bones and blood to foretell the future was also practiced. In Shakespeare's *Julius Caesar,* Calpurnia, Caesar's wife, begs him not to go to the senate because she has had a chicken's entrails read and the forecast was not good.

Many ancient cultures had beliefs about food and beverages that influenced their service. These ranged from strict taboos to customary practices.

The Development of European Haute Cuisine

During the Middle Ages in Europe, from the sixth to the fourteenth century, dining and culture in general progressed little and, in some respects, regressed. Still, inns and hostelries continued to serve travelers. In one publication from the period, we learn that inns offered three levels of service according to one's ability to pay. Monasteries took in and fed travelers on their way to the Holy Land. Public life revolved around the Church, which often sponsored community feast days. Markets offered food and drink for street consumption.

With the beginning of the Renaissance in the fifteenth century, the great flourishing of art, music, and architecture helped foster an environment in which dining and service, too, became more elaborate and sophisticated. Artisans and skilled tradesmen formed **guilds** to help regulate the production and sale of their goods. Several guilds involving food professionals—*Chaine de Rotessiers* (roasters of meat), *Chaine de Traiteurs* (caterers), *Chaine de Patissiers* (pastry makers)—grew in number and power until they effectively restricted their market.

During the fifteenth and sixteenth centuries, as more people ascended from poverty, the demand for better service and cuisine rose, especially in Italy. Books on social and dining etiquette appeared. In 1474, Bartolomeo de Sacchi, also known as Platina da Cremona, wrote a book on acceptable behavior while dining, dining room decoration, and good living in general. Soon after, the book *Il Cortegiano* (*The Courtier*) by Baldassare di Castiglione, became widely accepted throughout Europe as the official manual of behavior and etiquette. In 1554, Giovanni della Casa, a bishop who later was named Italy's secretary of state by Pope Paul IV, published *Il Galateo.* It was widely read and followed in its time as a guide to desirable conduct in society, and it became a classic of upper-class tastes of the European Renaissance. These last two works formed the foundation of hospitality service.

The Gastronomic Influence of Catherine de Medici

In 1533, the future king of France, Henry II, married **Catherine de Medici,** a member of one of Europe's richest and most powerful families. When she moved to France from her home in Florence, Italy, Catherine was shocked at the inferior level of food preparation and service. Even the French court and nobility ate common stews, soup, and roasted meats. Food was brought to the table in large pots or on platters, and diners helped themselves, dishing liquids up with ladles and picking solid foods up with their hands. They ate from wooden trenchers; daggers were their only eating utensil. Liquid in the trenchers was sopped up by bread and the solids scooped up by hand. Bones and waste were thrown on the floor to be picked up by household dogs and cats.

Catherine brought a staff of master cooks and servers to her new home. Tablecloths and napkins went on the tables, and the crude dishware and trenchers were replaced with fine dishes and carved goblets made of silver and gold. She introduced French society to knives, forks, and spoons, which the Florentines had been using since they were introduced to them by a Byzantine princess in the tenth century. The foods now were sumptuous and refined, and the service was lavish and elegant.

The French court and nobility quickly adapted to the new regime, and began to imitate it. Because the use of eating utensils was so new, those who entertained did not own many, and guests were expected to bring their own.

Fortunately, the king's nephew, who would later become King Henry IV and an enthusiastic gourmet, approved heartily of his aunt Catherine's standards. When he ascended the throne, he too required the highest levels of service at court level. France's nobility became connoisseurs of fine food, drink, and service. Upper-class standards continued to rise until formal dining reached lavish and elegant levels during the reigns of Louis XIII to XVI in the 1600s and 1700s.

The great majority of Europeans who were not members of the court, the nobility, or the privileged classes continued to eat and drink simply. They ate meals primarily at home, though inns and taverns catered to travelers and continued as gathering places for people.

The Restorante

In 1765, a Parisian named **Boulanger** opened the first restorante on the Rue des Poulies. Above the door was a sign in Latin reading, "Venite ad me ownes qui stomacho laboratis et ego vos restaurabo." (Come to me you whose stomachs labor and I will restore you.) Boulanger claimed the soups and breads he served were healthful, easy to digest, and could restore people's energy; hence, the name ***restorante.***

The guilds objected, claiming that only they had the right to prepare and serve food to the public. They sued Boulanger to stop him legally. Boulanger countersued and started a campaign to gain publicity. He had friends in high places that supported him. Soon he made his case a celebrated cause, even getting the Assembly and King Louis XV into the controversy. Boulanger won his suit. He protected his right to compete with the guilds, and opened the door for others to start similar operations. Soon *restorantes* opened in Paris and other cities in Europe, and the foodservice industry began.

Coffee was introduced to Europe in the seventeenth century. This brought about the development of the coffeehouse where coffee was served along with other beverages and some light food. Coffeehouses became popular as social gathering places for local people and acted as places where people could discuss common affairs and gain the latest news. They quickly spread all over Europe.

The **French Revolution** (1789–1799) ended the rule of the kings. Many of noble, wealthy, and influential people were killed or fled France. A new class arose, composed of artisans, capitalists, merchants, and intellectuals. This new middle class began patronizing restaurants, and the public demand for high-quality food, drink, and service increased. At the same time, many highly skilled cooks and servers who previously had served the upperclass found jobs in the new foodservice industry.

By 1805, only six years after the Revolution, fifteen fine-dining restaurants could be found in the area of the Palais Royal alone, serving the nouveau riche (new rich) the finest food with the best service.

Discriminating Gourmets

As this new class grew in stature, a group of discriminating gourmets appeared, and a number of them began to write about the art of fine dining. The French statesman Brillat-Savarin wrote *The Physiology of Taste.* Gimrod de la Reyniere edited the first gourmet magazine. Vicomte de Chateaubriand wrote many authoritative works on fine dining, and Alexandre Dumas père (father, or senior) compiled his classic *Grand Dictionaire de Cuisine.*

At the same time a group of chefs developed who also were interested in a high level of cuisine and service. The first of these was **Marie-Antoine Carême,** who trained a large number of very famous chefs to follow him and continue his high level of food service. They not only invented new dishes and new service, but also established rules on what foods should be served together, when they should be served during the meal, and the manner of service. Thus, it was Carême who first said that a heavy meal should be accompanied by a light soup such as a consommé, and a light meal should be accompanied by a heavy soup, such as a hearty lentil purée. Grimrod de la Reyniere later voted his approval by writing, "A meal should begin with a soup that, like the prelude to an opera or a porch to a house, gives promise of what is to follow."

The Growth of Service in Modern Times

The development of service after 1900 revolves around the tremendous growth of the foodservice industry—a direct result of increased industrialization, mobility, and disposable income. Today, one-fourth of all meals eaten in a day are consumed away from home. This represents 42 percent of the total dollars Americans spend for food and drink. People are eating out often and are demanding high-quality, yet increasingly casual service.

The Rise of Hotels

Greater mobility led to the growth of hotels and motels, which, in turn, affected food service. Luxury hotels were built to serve affluent patrons. One of the first of these was Low's Grand Hotel, built in London in 1774. It had more than 100 rooms and extensive stables for horses and carriages. It soon had many imitators throughout

FIGURE 1.2

Boston's Union Oyster House is the oldest continuous service restaurant in the United States. Courtesy of the Union Oyster House, Boston, MA.

England and Europe. Tremont House, which opened in Boston, Massachusetts, in 1829, was the first luxury hotel in the United States. The four-story building had 170 rooms with two bathrooms on each floor with running water. For the first time, guests could stay in their own rooms with their own key, all for $2 a night.

As railroads developed, hotels sprang up in every place with enough patronage to support them. New York City had eight in 1818; in 1846 there were more than a hundred. Chicago had more than 150.

The marriage of fine hotels, fine dining, and fine service culminated in the partnership of **César Ritz,** a hotelier, and **Auguste Escoffier,** one of history's greatest chefs. Ritz oversaw the front of the house and hotel management, while Escoffier saw to the kitchen and dining services. They made an unmatchable team; both had the highest standards. Ritz strove for elegant and luxurious service and spared nothing for the comfort and enjoyment of guests. Escoffier adapted and simplified the elaborate classic menus of his time to highlight top-quality cuisine and service. The wealthiest members of English and European society were their guests.

Ritz and Escoffier soon had many imitators. In the United States, a number of fine hotels appeared, such as New York's Astor House and Waldorf-Astoria, Chicago's

Palmer House, San Francisco's Palace and St. Francis Hotels, the Silver Palace in Denver, and the Butler Hotel in Seattle. The grand balls, banquets, dinners, and social affairs held in these urban hubs displayed the finest in elaborate socializing.

In 2005, more than 46,000 lodging properties, with 3.8 million rooms, have been built to accommodate American travelers, diners, and trade and professional events.

Restaurants and Service in the United States 

The first taverns in the United States were patterned after those in England. Their number increased throughout the seventeenth and eighteenth centuries, and they became an essential part of early colonists' lives. In 1656, the Massachusetts Commonwealth passed a law requiring every town to have at least one tavern. Not only did taverns provide food and drink, but they served as meeting places for people to discuss events and get the latest news. Inns also were established about the same time as taverns. They came into being largely to serve travelers and were located on the main travel routes.

The first recorded restaurant in the United States, The Exchange Buffet, a self-service, cafeteria-type operation, was built in the early 1800s opposite the current New York Stock Exchange. Boston's Union Oyster House, still in operation, opened in 1826. Delmonico's, Sans Souci, and Niblo's Garden were other fine eating establishments.

In the nineteenth century, dining out was restricted largely to the wealthy and to travelers. Around 1900, as the United States industrialized, workers began eating away from home more often. Cafeterias and lunch counters sprang up to serve both blue-collar laborers and white-collar professionals. As cities grew, shoppers and others were customers for cafes, coffee shops, family restaurants, and cafeterias. Institutional food service grew as well. The federal government mandated lunches in public schools in 1946. Dining out became a common experience.

According to the National Restaurant Association, today the foodservice industry is one of this country's largest industries, numbering 900,000 eating places doing approximately $437 billion in sales. This places it among the top ten industries in America in numbers of units and sales. It also employs more people than any other industry, a large number of whom are servers performing an essential and important service, without which this industry—and economy—could not exist. Each year, hundreds of thousands of people are needed to fill the demand for highly qualified and well-trained servers and managers.

FIGURE 1.3

Professional and well-trained staff are essential to the dining experience.

CHAPTER SUMMARY

Although the public service of food and drink began as a rather crude craft, over the centuries it grew until, by the end of the Roman Empire, it had reached fairly high professional levels. Restaurants, inns, hostelries, and other services had developed to allow the public to eat out.

Not until the tenth century did Europeans use eating utensils, although people living in the Middle East and Asia used utensils since the sixth century B.C.

Since ancient times, eating and drinking have played important parts in public gatherings and celebrations. Every culture has established service and culinary customs based on their religious beliefs and physical environment.

The Renaissance ushered in an era of fine dining in Europe limited to noble and royal families. Catherine de Medici changed French eating when she became their queen, starting the growth of dining standards that reached lavish and elegant standards. Many of these standards were taken over by restorantes, which started just before the French Revolution and were open to the public. Standards for fine dining also were set with the writings of a number of great French gourmets. These standards developed in France, influencing eating all over Europe and other parts of the world.

After the French Revolution, a new middle class arose with adequate incomes to eat out. Often, such dining was more casual but much fine dining still existed. Restaurants thrived.

Both ancient customs and modern values dictate that guests should be treated well by their hosts, and that hosts should make every effort to see that their guests come to no harm.

Servers who approach their work professionally are able to deliver exceptional service. This entails anticipating guests' needs and wants and exceeding their expectations.

Related Internet Sites

Food and Nutrition Information Center

This Site provides the principles that the food service industry has utilized to fight food-borne illnesses.

www.nal.usda.gov/fnic/pubs_and_db.html

History of Foodservice and its Role Today

From early kitchen productions to menus all around the modern world, this site enables you to understand the different aspects of food service at its best.

www.schonwalder.org/Menu_1_Iq.htm

Restaurant and Hotel Services

Research and input from industry professionals

www.restaurantedge.com

Prohibition of Alcohol

www.prohibition.org

Key Terms

Boulanger
Marie-Antoine Carême
Auguste Escoffier
French Revolution
guilds
Halal
Kosher
Catherine de Medici
restorante
César Ritz
sanctuary
taberna vinaria
thermopolium

CHAPTER REVIEW

1. What part did early Christian churches, monasteries, and convents play in promoting travel and service in Europe?
2. What was a common reason for ancient Greek feasts?
3. How have religious symbols and traditions influenced food?
4. What were Catherine de Medici's contributions to French dining?
5. What was a guild? What control did a guild have?
6. Who was Boulanger? What was his contribution to food service?
7. What were some of the special contributions of the gourmet chefs of the 1800s?
8. How did industrialization contribute to the U.S. foodservice industry?
9. How has the concept of sanctuary affected modern notions of hospitality?
10. Why is it important to exceed guests' service expectations?

CASE STUDIES

Excavating Ancient Ruins

You have taken a job with a company hired by the Turkish government to excavate some ruins of an ancient city in Turkistan. The company comes across evidence of a large communal kitchen and dining area. What would you hope to find there that would tell one much about this ancient culture's foods, methods of preparation, dining methods and cultural dining practices, etiquette customs, and social dining practices?

Ancient Foods

Ancient peoples did not have the food resources we have today. What do you think they ate? Do you think it may have had other meanings to these cultures than just being something to sustain life? Dig up information about this on the Internet and other sources. For instance, the novel *Quo Vadis* by Henryh SienKiewicz and W.S. Kuniczak includes a marvelous description of a Roman banquet. Books on the history of Rome will have descriptions of the feeding of the masses from the grain stores of the Roman government.

THE PROFESSIONAL SERVER

OUTLINE

I. Introduction
II. Finding Work
  A. The Interview
  B. Interviewing
  C. The Legal Side of Hiring
III. Looking Professional
  A. Uniforms
IV. Demeanor and Attitude of Successful Servers
  A. Delivering What Customers Expect: Operation Knowledge
  B. Maintaining a Positive Attitude
  C. Courtesy
  D. Tact
  E. Sincerity and Honesty
  F. Camaraderie
V. Learning Skills
VI. Product Knowledge
VII. Suggestive Selling
VIII. Organization
IX. Tips
X. Unions
XI. Laws Affecting Servers
  A. Privacy Act
  B. Fair Labor Standards Act
  C. Family And Medical Leave Act
  D. Civil Rights Act
  E. Beverage Alcohol
  F. Immigration Reform and Control Act
  G. Americans with Disabilities Act
XII. Chapter Summary
XIII. Chapter Review
IX. Case Studies

LEARNING OBJECTIVES

After reading this chapter, you should be able to:

- Outline the skills and behaviors common to professional servers.
- Explain how hiring qualified employees helps an operation deliver excellent customer service.
- Describe laws that affect employees and employers.

Introduction

This chapter looks at the process of finding work as a server, and then at the factors that make a server professional in the eyes of guests, colleagues, and employers. These factors include a professional, neat appearance; a positive and helpful demeanor; courtesy and tact; a high level of knowledge of food and service; and the ability to suggest menu items and stay organized throughout the workday. This chapter provides an overview of these topics. In addition, it covers some of the basics of tips and unions, as well as the laws that affect servers.

Finding Work

Finding a good server position is not always easy. There might be many job openings, but it might be a challenge to find one that is a good match with your talents and availability. Word of mouth is one good way to find a position. This gives the server first-hand information from someone who knows the operation. Other sources of job openings include help-wanted advertisements, union headquarters, employment agencies, and just walking into operations.

The Interview

Remember when going to an interview to present yourself as a desirable employee. Do not brag but present yourself in a positive manner. Precisely explain your qualifications. Remember that body language tells the interviewer a lot. Display your confidence. Show the interviewer the same professional behavior you will use on the job. Dress and grooming should be neat, simple, and appropriate to the operation.

As a manager, hold interviews only when you are seriously considering an individual for a position. Interviews are designed to gather information about skills, personality, and job knowledge. Often a review of the application leads to interview questions. A broad amount of information is obtained by asking open-ended questions, such as, "You have worked at several places during the last year. Why did you leave each one of them?" This can lead to further questions that might give some revealing information about the applicant. A server may go through a preliminary interview with the manager or supervisor of the operation. The supervisor who is to immediately oversee the work of the server should be the final interviewer.

Interviewing

As a manager in the competitive foodservice industry, you will have the greatest success only if you actively strive to hire and retain the most qualified, talented, and motivated employees. The process begins with screening the most qualified and appropriate job applicants, identifying outstanding job candidates through effective interviewing, and selecting employees who are likely to remain and develop within your operation.

Most interviews will be divided into four parts: (1) preparation; (2) the interview; (3) ending the interview; and (4) evaluation. The key to conducting effective employee interviews is to plan your part of the interview in advance. By preparing interview questions and structuring the interview's direction before the meeting, you are much more assured of getting valuable information about the candidate than by simply winging it.

Most operations will find it helpful to establish **job description specifications,** which are written definitions of the requirements of the job and the person that should be hired to fill these requirements. It takes a person who knows the job and what it entails to write these. The job specification lists knowledge, skills and abilities, work experience, and education and training. Write specifications based on the information needed to perform job duties; the ability to perform a task, or behave in a certain way; and any specific skills, if applicable.

Another item that should be done before interviewing is to set up questions to use to get the information needed to judge the candidates' suitability for the job. These should cover the areas of: (1) education; (2) motivation; (3) ability to work with others; and (4) relevant personal characteristics. See **Exhibit 2.1** for some examples of questions that might be asked in these four areas.

Open-ended questions require more than a yes or no answer, and they encourage candidates to talk about themselves and their experiences. By asking open-ended questions the interviewer gets a chance to get much more desirable information about the candidate that might be helpful in estimating the candidate's suitability for the position. Listen carefully to the candidate's answers. In fact, you should spend most of your time in the actual interview listening to the candidate.

There are other things one needs to do in planning for the interview. The room in which the interview occurs should be private, orderly, and unintimidating. Arrange to sit next to the applicant. Do not have anything such as a

EXHIBIT 2.1 Some Open Questions to Ask in Interviewing

Education

1. Who was your favorite teacher? Why did you like her or him?
2. What courses did you take? In which did you excel?
3. In what extracurricular activities did you participate?

Motivation

1. I see you have had previous experience as a server at ______. What were your job responsibilities? Did you enjoy doing them?
2. What are your professional goals? What do you want to be doing one year from now? Five years from now?
3. What do you expect from supervisors? From co-workers? From buspersons or others who work for you?
4. What did you like about your contacts with the kitchen crew? What did you not like?

Ability to Work with Others

1. What advantages do you see in working with others?
2. Describe some unpleasant experiences you have had with co-workers and how you handled them.
3. If you were given an inexperienced person to work with and train, what would you do to help this person learn, while at the same time doing the required work?

Relevant Personal Characteristics

1. When you are a guest in a restaurant, what type of service do you expect?
2. If you saw a co-worker stealing, what would you do?

desk or table between you and the candidate, in order to help put the candidate at ease. Have materials ready to give the applicant. Notify others who will be interviewing the person. Make arrangements so you will not be interrupted during the interview. Have a note pad handy to use in taking notes.

One must remember in conducting the interview that the interviewer is also under scrutiny, and that the applicant is often making an evaluation of whether he or she wants to work there. The interviewer should greet the applicant warmly and be pleasant during the entire interview. Body language, such as facing the person who is talking and maintaining eye contact, also should be positive. Listen actively by nodding, maintaining eye contact, asking questions, and at times repeating what the candidate is saying. The only way you are going to find out what you want to know is to listen and hear it from the candidate. Of course, there are other factors to note, such as dress, the way the candidate conducts himself, and so forth, but the main source is through what the candidate says. Short comments, such as "Yes, that's good" or "I see," also indicate active listening. Don't expect spontaneous answers. Let the candidate think out answers before giving them.

One should not mislead candidates or make false promises. State frankly what is good and what is bad about the position. It is much better to be honest with the candidate about the hard parts of the job.

Before ending the interview, give the candidate a tour of the facility, explaining things that are of interest about the job. Introduce the candidate to others in the operation. Be sure to ask the applicant if there is anything more that he or she desires to know before ending the interview.

In ending the interview, thank the candidate, and indicate what will be done to inform her or him of the decision. If a date and time is given, be sure to observe it. Go with candidates to the exit and wish them well. Even if the interview went poorly, still be positive and courteous. The fact that the candidate would take the time and trouble to come for an interview is worthy of polite, considerate treatment.

After the interview, quickly review your notes, adding anything that you might have wanted to jot down but did not have the time. Summarize your judgment as to the candidate's suitability for the position. Before making any decision, be sure to weigh all the facts. Today, with the shortage of labor, one is apt to make hasty decisions; avoid this. Interviewing and hiring good employees is crucial to a successful operation.

The Legal Side of Hiring

There are legal restrictions to observe in the screening, interviewing, and hiring of employees, and violators will find they face severe penalties for not observing them. Only the federal laws in this area are reviewed below, but managers and supervisors should also know and observe all state or local requirements.

The laws and their limitation affecting recruiting and hiring appear in **Exhibit 2.2.** At no time should one mention or ask the candidate to give information on any of the following:

- Race, religion, age, or gender
- Ethnic background
- Country of origin
- Former or maiden name or parents' name
- Marital status or information about spouse
- Children, child-care arrangements, pregnancy, or future plans to become pregnant
- Credit rating or other financial information, or ownership of cars or other property
- Health
- Membership in an organization
- Voter preference
- Weight, height, or any questions relating to appearance
- Languages spoken, unless the ability to speak other languages is required of the position
- Prior arrests (convictions are legal)

EXHIBIT 2.2 Federal Laws Affecting Recruiting and Hiring

Law	Description
Federal Insurance Contributions Act (FICA) (1937)	Source of federal payroll tax law, especially regarding Social Security
Fair Labor Standards Act (1938)	Establishes requirements for minimum wages, work time, overtime pay, equal pay, and child labor
Equal Pay Act (1963)	Requires employers to provide employees of both sexes equal pay for equal work
Civil Rights Act (1964)	Forbids discrimination in employment on the basis of race, color, religion, or national origin; sex and pregnancy are covered in the employment section
Age Discrimination in Employment Act (1967)	Prohibits discrimination against job applicants and employees over age 40
Equal Employment Opportunity Act (1972)	Prohibits discrimination based on race, color, religion, sex, or national origin (amended Civil Rights Act of 1964)
Vietnam Era Veterans Readjustment Act (1974)	Protects Vietnam veterans from any job-related discrimination
Immigration Reform and Control Act (IRCA) (1986)	Forbids employers from knowingly hiring anyone not legally authorized to work in the United States
Americans with Disabilities Act (1990)	Prohibits discrimination against qualified individuals in employment; requires employers to make reasonable adjustments in facilities and practices to permit participation of disabled persons

The key is to ask only job-related questions. Questions such as, "Can you work nights and weekends?" and "Are you available to work overtime when needed?" are appropriate. Careful planning enables the interviewer to ask questions that are both legal and effective.

Looking Professional

Servers who wear crisp, clean uniforms and are properly groomed will make a positive first impression on guests, their employers, and colleagues. This typically translates into higher tips, better shifts and table sections, and enhanced opportunities for recognition and advancement.

FIGURE 2.1

A well-dressed, well-groomed server makes a positive impression to guests. Courtesy Corbis Digital Stock

Uniforms

Styles of uniforms vary greatly from operation to operation. Many bars, family-style restaurants, and casual-theme operations feature servers in khaki pants, shorts, T-shirts, cotton button-downs, or polo shirts. Servers in upscale fine-dining restaurants often wear ties and black aprons over formal clothing. Institutional food service uniforms typically resemble traditional uniforms rather than casual street clothes. No matter what an operation's server uniform, several things are consistent. For instance, the uniform should be clean, wrinkle-free, and well-fitting. Shoes should be comfortable and sturdy enough to withstand hours of standing, walking, and direct contact with slippery surfaces.

If servers carry pens, order pads, corkscrews, or matches, the items should be kept in pockets below the waist to prevent them from dropping. Never put a pen or pencil in your mouth or behind an ear. Extra uniforms should be on hand for emergencies. Jewelry should be limited to post earrings, rings, and watches, since anything dangling could end up on or near guests' food.

Although personal hygiene may be a sensitive subject, it is vital to food safety and proper server presentation. Unclean servers can disturb guests, and even spread illness or other contamination. Servers should bathe and wash hair frequently. Clean clothes should be worn. It is preferable to have workers change into clean clothes at the workplace. If unable to do this, workers should come from home directly to work. Servers are on their feet almost all their shift, so comfortable shoes should be worn—no platform, high-heeled, or open-toe shoes. Wear hair restraints. These are often required by health authorities. Nets, caps, or hats may be used. If hair is longer than collar length, it should be tied back or pinned up.

Basic hand care includes keeping nails short and clean. Wearing garish nail polish or artificial nails should not be permitted. Plastic gloves or using tongs should be required in handling food. All sores, cuts, or infections should be neatly covered by bandages or bandaids.

There are some sanitation do-nots in handling eating utensils:

1. Do not touch the inside of cups, glasses, silverware, or dishes where food or liquid or lips will touch.
2. Do not handle clean utensils without washing hands after clearing tables.
3. Do not leave trays used for serving with soil on them.
4. Do not use soiled napkins, serviettes, or other linens.

Demeanor and Attitude of Successful Servers

A pleasant demeanor and positive attitude have as much or more to do with success in serving as knowing how to do the work. In fact, a survey of managers to find out the cause of servers' failure on the job indicated that only 10 percent were discharged because they did not know how to do the job. The other 90 percent were discharged largely because of personal traits or negative attitudes.

Delivering What Customers Expect: Operation Knowledge

To serve guests well and answer their questions intelligently, servers must be informed about the company for which they work and the products offered. Employees should know the following:

- Days and hours of operation
- History and background of the company
- Theme and concept of the decor
- Names of managers, management assistants, and supervisors
- Places of interest in the local area

Maintaining a Positive Attitude

One of the most important personality traits a good server must possess is a positive attitude toward work, colleagues, and serving the public. Good servers believe they can deliver and try not to dwell on failures, but correct their mistakes and learn from them.

Having a positive attitude toward work allows the server to make progress in learning and to develop increased proficiency. Setting small goals and achieving them builds confidence. After this, more difficult goals appear within reach. It is important to be prepared to take advantage of opportunities when they appear.

Making an earnest effort to be friendly with guests and to please them brings with it both financial and professional rewards. Servers should try to develop challenges for themselves, such as trying to win over difficult guests. **Exhibit 2.3** contains a list of the fundamentals of professional service.

Servers should never talk about personal problems or inappropriate topics with guests or within guests' hearing.

EXHIBIT 2.3 Professional Service Fundamentals

Adhering to these standards ensures professionalism in manner and service.

- Be sure your personal style matches the style (formal, casual) of the operation.
- Don't initiate a superfluous conversation unless at the request of the guest.
- Humor can be positive and pleasing to the guest when properly applied.
- Timing is of crucial essence. For instance, if a guest orders a glass of wine served with the entree, and it comes even five minutes after, it can annoy the guest.
- Little things mean a lot. Everything that can be done to make a guest's experience more enjoyable, comfortable, and easy will always be appreciated.
- In a family restaurant, provide activities for children, such as crayons or puzzles, so parents can enjoy their meal.
- Always present the check face down, so that only the host(ess) can see the total.
- When saying goodbye, if the guest should extend the hand, the server should offer a firm handshake.
- Maintain proper eye contact, which is a sign of attentiveness and sincerity, at all times.
- Place dishes on the table gently.
- Guests should never be hurried, and should never be given the impression that they are being rushed so that others can be accommodated.
- In handling china and glassware, never touch the top or inside of a glass, or the surface or edges of a plate.
- To be a team player is a must. Help other servers whenever possible.

They should never complain in the dining room about the lack or paucity of a tip. In fact, it is not the server who should be the focus of attention at all, but the guest. Although this separation is certainly a challenge, support from colleagues and good cooperation can go far in keeping servers focused on their guests and giving great service.

Courtesy

In life, courtesy means being polite, gracious, and considerate toward others. In food service, it means putting the guest's needs before one's own. Respect for others and a willingness to help are key. Courtesy should be automatic and natural. It is displayed through words and actions. Being courteous does not mean being servile or fawning. Good servers are professionally courteous, showing a serious regard for their work. Even difficult guests, when treated courteously, will return the favor. The few who don't are rare, and servers should try to meet their needs and not take their mistreatment personally.

Tact

Tact is the art of saying and doing the right thing, using the right words at the right time. It is also an intuitive sense of what to do or say in order to maintain good relations and avoid offending guests. Behaving tactfully might be remembering and using guests' names, using diplomacy in adverse circumstances, or asking a guest to take a phone call when it is necessary to tell her that her credit card has been rejected. Being tactful means handling sensitive situations so that everyone involved is left with their dignity intact.

EXHIBIT 2.4 Excellent Service on the Job

Providing excellent service is a good way to establish loyal customers.

- Do not correct guests if they mispronounce item names.
- Anticipation is a fundamental component of service. Guests should not have to ask for refills on coffee, water, and so on.
- Even the most helpful service, given with improper timing, can be perceived as poor service.
- Only talk about yourself when asked. Guests are the celebrity at the meal.
- Never allow your emotions to get the best of you. The service you give must remain consistent and professional, especially when dealing with difficult guests.
- If a napkin or piece of flatware falls on the floor, replace it immediately with a fresh one.
- Before clearing something, ask if the guest is finished.
- Describe items in an appetizing manner, such as, "Our special, Southern Fried Chicken, comes from an old southern recipe using special herbs and spices; it is crisp outside and moist and tender inside."
- Too much zeal in serving can bother guests. This usually discourages tips. Service that brings in tips is pleasant, effective, and unobtrusive.
- Do not stand nearby when a guest is paying the bill. Most guests want privacy when figuring out a tip or counting out money. They may wish to discuss the tip without the server being present.

Sincerity and Honesty

Sincerity and honesty are shown by behaving naturally, and not in a forced or phony way, toward guests. A forced smile and "canned" lines ("Have a nice day") are obvious clues to insincerity. Being pleasant while serving is really all that is needed.

Being frank and telling the truth are important. Servers who make mistakes should simply admit the mistake and correct it as quickly as possible. Guests will appreciate the forthrightness and the effort. Excellent service will be easy to perform if servers follow the tips in **Exhibit 2.4.** And in **Exhibit 2.5** listed are some helpful phrases that servers can use.

EXHIBIT 2.5 Helpful Phrases

Pleasant, courteous phrases are always appropriate when dealing with customers.

- Good evening (morning, afternoon), and welcome to ____.
- My name is ____, and I'll be your server this evening (morning, afternoon). If there is anything that I can get for you, please let me know.
- May I take your order now, or would you prefer a little extra time to go over the menu?
- Do you have any questions about our menu?
- May I suggest a wine to complement your entree (or coffee with your dessert)?
- How is everything? Is the ____ done to your order?
- Thank you. It's been a pleasure serving you. We look forward to your return.
- I hope to see you again soon, Mr. ____ or Ms. ____.

Camaraderie

Camaraderie is the ability to get along with people. When team relationships falter, guests suffer. No matter how you serve guests in the foodservice industry, your ability to work with others to serve them well will help you move ahead professionally.

Learning Skills

Professional servers must learn continually throughout their jobs or careers. Learning and training in service skills is accomplished in several ways: through DVDs and videotapes, study courses, computer programs, CD-ROM programs, simulations, online courses, training sessions, and other servers. No matter what the method, both the trainer and server are responsible for seeing that learning takes place and is put into practice on the job.

Product Knowledge

Just as a doctor knows the human body and the mechanic knows cars, servers must know about the products they serve. If a guest asks about a menu item, the server should provide all possible answers. Servers should study their menu to know how items are prepared and what they contain, and know all specials before a shift begins. This increases a server's opportunities for suggestive selling and increasing check averages and tips.

In addition to basic knowledge about the operation, servers must know a good deal about the menu. In the meeting of servers before the meal, servers should taste unusual dishes on the menu and know ingredients and methods of preparation of menu items. Know the following:

- What items are on the menu
- Signature items

- Promotional items
- Specials
- Estimated runout time
- Items of interest: For example, oysters flown in from the west that morning

Suggestive Selling

Suggestive selling involves the extremely important role your servers have in suggesting items to guests, selling individual menu items, and increasing check averages and tips. Suggestive selling involves offering all of your guests the full range of products and services available in your operation. The more guests know about what menu items are available, how they are prepared, and why your operation is pleased to offer them, the more likely they are to enjoy every aspect of their meal. Suggesting menu items benefits your servers in increased tips, guests in increased enjoyment, and the operation in increased profits.

A knowledge of menu terminology is also essential. The best servers not only know what menu terms mean but also the explanation of the terminology. Knowing menu items can be a matter of pride and accomplishment in the profession of serving.

Inexperienced servers may not know many menu terms. The way to learn is to take a menu and ask the chef or manager to explain what is unfamiliar. Usually, in the line-up session before the meal, servers are told what menu items are.

Organization

Organizing one's own work and time is essential. A disorganized server will have trouble with timing, such as knowing when to take an order, pick up items, or present the check. Disorganization breeds a frantic pace, tension and nervousness, and frequent attempts to catch up. Good servers have a rhythm.

Good organization will also help you anticipate guests' needs. Good observation is one of the most important factors in organization and timing.

Tips

The meaning of tips—**T**o **I**nsure **P**rompt **S**ervice—is important to remember. Many servers take tips for granted, and make little or no effort to do additional work to earn one. This may be the fault of guests who tip 15 to 20 percent of the bill regardless of the quality of the service, so that servers know that extra effort is really not needed.

However, with guests becoming more service and value minded, more discrimination is being shown in giving tips. Servers are expected to give good service, and in return, can expect a fair tip.

Some servers far outdistance others in earning tips. This comes through actions and words. They always welcome guests and thank them, and give the extra effort needed to please guests. **Exhibit 2.6** outlines the basics every server should know in order to please guests.

EXHIBIT 2.6 Basics of Good Service

Be sure to remember these service tasks when attending to guests' needs.

- Serve items from the guest's left side.
- Remove items from the guest's right side.
- Serve and remove beverages from the guest's right side.
- Serve main food items at the six o'clock position.
- Handle fine glassware by the stem.
- Keep water glasses two-thirds full.
- After filling a tray or bus pan with soiled and leftover food, cover it with a napkin.
- Serve all bar beverages with a napkin or coaster.

Tips are considered part of a server's salary, and must be reported as such to the Internal Revenue Service. An employer is required to report 8 percent tips for each server in an establishment. This is calculated by dividing the total sales check amount by the number of guests served.

Unions

Working in a **unionized operation** means that servers are represented to management by union representatives. In a unionized operation, the employer company typically signs a contract with union representatives covering job classifications, job duties, scheduling, pay, grievance procedures, vacation time, length of work week, break times, sick leave, termination, and so on. Unions charge their members dues to sustain their operation.

It is advisable for an operation's manager to have job descriptions and job specifications written before signing a labor contract if he or she does not want to forfeit that function to the union.

Union contracts often require that employees performing unsatisfactorily be warned orally and in writing a specified number of times before termination. This also happens to be a very wise management policy, since it protects managers in the case of wrongful termination lawsuit. Warnings should state a specific cause and incident description, and a description of how performance is expected to improve. Employees must be given a chance to correct their actions within a certain time period.

Managers should make every effort to settle employee matters internally. However, if a worker has a grievance which is not satisfied, he or she can contact the union and have it take up the matter with the company management. If the union believes the company has violated a contract term, it contacts the company to settle the matter amicably. If the matter is not settled, the union may appeal to a grievance committee to hear the case and make a decision. A union contract usually contains an agreement by both the union and management that the committee's decision will be binding on both parties. If not, the employee or the company could appeal to a court for a decision. Grievance committees are limited to hearing cases that arise within a specified area.

Laws Affecting Servers

A number of laws relating to hiring and work affect both employers and servers. The most important of these follow.

Privacy Act

The **Privacy Act of 1974** forbids employers from asking non-job-related questions that might discriminate against a group of qualified job applicants. The act applies not only to interviewing potential employees but also to discussing matters with current employees.

Fair Labor Standards Act

The **Fair Labor Standards Act** of 1938 protects workers between ages 40 and 70 from discharge because of age. This act also covers teenage workers, working hours, and union activities.

Family and Medical Leave Act

The **Family and Medical Leave Act** (FMLA) of 1993 requires employers with 50 or more employees to offer up to twelve weeks of unpaid leave in any twelve-month period for any of the following reasons:

- Birth, adoption, or foster care of a child
- Care for a child, dependent, spouse, or parent with a serious health condition
- Care for the employee's own serious health condition

The act includes some other provisions for both employer and employee that guide their relations dealing with family and medical leaves.

Civil Rights Act

Discrimination against job applicants and employees is prohibited in Title VII of the **Civil Rights Act** of 1964 and 1991. The act is administered and enforced by the **Equal Employment Opportunity Commission (EEOC).** It is unlawful to "fail or refuse to hire or discharge any individual or otherwise discriminate against any individual" on the basis of race, color, religion, sex, or national origin. The reason for not hiring an applicant must be job-related. For instance, in New York City a group of restaurants refused to hire women as captains and servers, saying that males only would be accepted by their guests. The EEOC ruled against them. In some cases, foodservice operations must train applicants to become suited to the work.

The Civil Rights Act further covers wrongful discharge on the basis of age, disability, and participation in collective bargaining or union activities. The penalty to the employer may be reemployment of the fired employee, court costs, attorney's fees, and in some claims penalties for committing an act against public policy or outrageous employer conduct.

The act now also covers sexual harassment on the job, which is defined as "unwelcome sexual advances, requests for sexual favors, and other verbal or physical conduct of a sexual nature . . . when:

1. Submission to such conduct is made either explicitly or implicitly a term or condition of a person's employment; or,
2. Submission to or rejection of such conduct . . . is used as the basis of employment decisions, affecting such person; or,
3. Such conduct has the purpose or effect of unreasonably interfering with a person's work performance or creating an intimidating, hostile, or offensive working environment."

FIGURE 2.2
Knowledge of the menu by the server allows for suggestive selling, which will also likely increase check averages and tips. Courtesy PhotoDisc, Inc.

A majority of cases on sexual harassment have favored the plaintiff and imposed substantial penalties on offenders. Judgments against defendant employers can be substantial.

Beverage Alcohol

Sale of alcohol is regulated by state and local laws. They generally cover licensing, permits, and how liquor may be sold. Some states called **control states** handle the sale and distribution, while others permit retail beverage establishments to purchase from prescribed dealers. They also typically address how to handle disruptive patrons. It is wise to call in the police to help handle a difficult guest.

Dram shop laws hold anyone serving alcohol liable if an intoxicated patron injures or kills a third person; hence, the equivalent term *third-party liability.* Courts may grant large damages against establishments and servers for violating dram shop laws. Insurance companies faced with paying such penalties have raised insurance fees to cover these costs.

To reduce such liability, many jurisdictions require that those serving alcohol be trained in serving it responsibly. Responsible service includes never serving alcohol to three types of people: intoxicated patrons, minors, and known or habitual alcoholics. Detecting a minor can be difficult. An operation should have specific steps in place for verifying identification. Documents without pictures or dates should not be accepted. If in doubt, examine the documents carefully. At times, minors accompany older persons who buy them beverage alcohol. Some operations or laws may prohibit minors from entering the premises of an operation selling beverage alcohol.

Detecting intoxicated people and alcoholics also is difficult. Servers must be trained to observe and interpret guests' behavior, and monitor guests' drinking. If an intoxicated guest tries to drive away, a server or manager should call the guest a cab, or, if all else fails, call the police.

Immigration Reform and Control Act

The **Immigration Reform and Control Act** (IRCA) of 1986 makes it illegal to hire aliens not authorized to work in this country. Employers must verify citizenship at the time of hiring. If asked before hiring (during the interview), nonhired applicants can claim discrimination. There is a grandfather clause that says that aliens hired before November 7, 1986, can be kept on the job. In addition to these, some aliens may be given a permit to work in this country. Every new employer must complete a Form 1–9 upon hiring and put it into the employee's file; its purpose is to verify that the new employee has submitted satisfactory proof of identity and work authorization, if the latter is required.

Americans with Disabilities Act

The Americans with Disabilities Act prohibits discrimination against people with differing levels of ability, requires reasonable accommodation for employees with disabilities, and requires public places and services to accommodate guests with disabilities. This means that not only is it good business to accommodate all guests, regardless of ability, but it is also the law.

CHAPTER SUMMARY

Looking professional is key to a server giving a good first impression. When looking for work, word of mouth can be a reliable source, as are newspaper advertisements, employment agencies, and walking into operations. The art of applying and interviewing for work is very important in finding desirable positions.

Demeanor and a positive attitude are crucial. Servers must be sincere, honest, and courteous. Good teamwork helps give better service. It is important for servers to know menu terms and how items are prepared.

Some servers are able to make more tips than others. Much of their success comes from knowing how to please guests. Tips are considered salary and must be reported as such to the IRS.

Unions sometimes represent employees in dealing with management.

Certain laws have an influence on how servers get work and how they perform it. The Privacy Act protects candidates and employees from inappropriate inquiries. The Fair Labor Standards Act protects servers from the ages of 40 to 70 from discharge because of age and has sections governing the work of teenagers. The Family and Medical Leave Act requires employers with 50 or more employees to offer up to twelve weeks of unpaid leave in any twelve-month period for reasons related to family and personal health.

The Civil Rights Act bars discrimination against employees because of race, color, religion, sex, or national origin. The Equal Employment Opportunity Commission (EEOC) enforces the act, which also covers cases involving sexual harassment on the job.

State and local dram shop laws create third-party liability on people serving beverage alcohol. The laws hold servers responsible to third parties injured or killed by intoxicated patrons, and encourage servers to be trained in monitoring alcohol service to guests.

The Immigration Reform and Control Act holds employers responsible for verifying the legal working status of employees. With a few exceptions, illegal aliens are prohibited from working in the United States.

RELATED INTERNET SITES

Tips on how to be a professional server
www.soyouwanna.com/site/syws/waiter/waiter3.html

KEY TERMS

Civil Rights Act
contol states
dram shop laws
Equal Employment Opportunity Commission (EEOC)
Fair Labor Standards Act
Family and Medical Leave Act
Immigration Reform and Control Act
job description specifications
Privacy Act of 1974
unionized operation

CHAPTER REVIEW

1. What is a grievance committee, and of what use can it be to a server?
2. Is the question, "Do you have any disabilities?" appropriate to ask in a job interview?
3. What should servers do to maximize tips?
4. What constitutes sexual harassment?
5. What is third-party liability?
6. Describe a typical server's uniform for a casual operation.
7. How does being tactful help a server when a guest's credit card has been denied?
8. What do job specifications contain?
9. Describe some ways to show you are listening actively during an interview.
10. Is it appropriate to ask a female candidate if she is planning to have a baby anytime soon?

CASE STUDIES

Establishing Job Descriptions

You are the manager of a family restaurant and want to establish job descriptions for the various jobs on the staff. In the service area you will need descriptions for the head waiter, servers, cashier, and hostess. Write these job descriptions for the service staff.

Unwanted Suitor

Christine works nights at Charlie's restaurant while going to college during the day. In three years of attending the university in the day time she has earned her B.S. degree. She is continuing to work at Charlie's while earning her master's degree. Things are proceeding well until a young man, whose first name is Bob, is hired as an evening server. He is a bit older than normal college age and is attending the same college as Christine while working on a business degree. He is attracted to Christine, who at first accepts his attentions. They have several dates together, but it turns out that Bob is more interested in her than she is in him. Christine tries to cut him off, but he is persistent and begins to annoy Christine so much she feels she can no longer work at Charlie's. However, she really doesn't want to leave. She has made the job fit her needs so well that she could never duplicate it elsewhere nor find a job that pays as well as her current job does. What should she do? How does she get rid of Bob? How would you as a manager help to resolve this situation.

Exceeding People's Needs

Outline

I. Introduction
II. Managing Guest Complaints
  A. Complaints as Opportunities
  B. Some Don'ts in Dealing with Angry Guests
  C. Taking Action
  D. Why Guests Complain
  E. Signs of an Unhappy Guest
  F. Avoiding Complaints
  G. Keeping Managers Informed
III. Serving Guests with Special Needs
  A. People with Disabilities
  B. Communication Issues
  C. People with Vision Impairment
  D. People with Special Dietary Needs
IV. Chapter Summary
V. Chapter Review
VI. Case Study

Learning Objectives

After reading this chapter, you should be able to:

- Explain the steps in resolving customer complaints in order to satisfy guests.
- Describe ways to manage service to customers with special needs.

Introduction

It is important for foodservice managers to create a work environment in which all employees are encouraged to try to satisfy and please guests. They must be trained to focus on guests at all times, and put their needs first. From the moment guests enter the operation to the time they leave, servers must make them feel comfortable, welcome, and anxious to return. By anticipating guests' needs, employees will be able to serve guests in the most efficient and effective way possible.

High-end foodservice operations appear to be meeting the challenge of providing customers with good service. According to the National Restaurant Association's Tableservice Restaurant Trends 2004, 74 percent of customers surveyed at higher-priced restaurants gave a good or excellent rating to the service received. However, regardless of any high service marks an establishment may receive, each customer complaint should always be taken seriously.

To avoid customer service problems, servers should never do the following:

- Ignore customers.
- Rush customers who wish to stay longer.
- Keep customers waiting without acknowledgment and explanation.
- Serve anything substandard.
- Allow customers to sit at a dirty table.
- Allow dirty or unsanitary conditions to exist.
- Behave rudely or even indifferently to customers.

Managing Guest Complaints

To solve a guest complaint, the server should ascertain what the complaint is, correct it if possible, and show the guest that the complaint is a serious matter to the operation. This will usually satisfy most customers. Be sure to offer a sincere apology for any inconvenience. Showing sincerity, concern, and a desire to correct the problem is the right approach.

Complaints as Opportunities

Often guests complain with good intentions. They may be complaining about something they think might cause others to complain and they are trying to call it to management's attention. Usually, this complaint is given in a calm, friendly, and suggestive manner. Management should be extremely grateful, promise to take care of the matter, and then do so.

Do not look at a complaint as an attack on anyone in the operation or how the operation is managed. Complaints can be helpful in improving business. Often, managers learn about problems only from their customers. It is not easy to attract new customers, so it is never wise to lose any. Receiving a complaint is much better than having a dissatisfied guest leave and never come back. Accepting the complaint allows one the chance to correct it. The silent, dissatisfied patron is a hazard to profitable business. Unhappy guests who do not come back often discourage others from coming. While word of mouth may be the best advertisement, it also can be the most destructive. **Exhibit 3.1** lists eight of the most frequent complaints made by customers.

Some Don'ts in Dealing with Angry Guests

- Don't be defensive.
- Don't assign blame.
- Don't make promises you can't keep.
- Don't keep phone guests on hold.
- Don't refuse a guest's request without an explanation.

Taking Action

No one appreciates being attacked by an irate customer. Servers must be trained to stay calm and composed in these situations, detached from personal negative feelings, and staying professional at all times. The rule in receiving a complaint is this: Listen, accept responsibility if it exists, then resolve the problem if you can.

The first thing a server should do is listen to the guest and find out the real cause of the complaint. Try not to give excuses, such as, "We're short three servers tonight." If you stay calm and listen, you might get a clue as to how to solve the situation. Ask questions until the complaint is understood. Restate it so the guest knows you understand the complaint correctly.

All guests want and deserve to be heard fully. Allowing them to have their say might make them feel better and even see that what they are angry about is not as important as they thought. Try to look at the situation from the guest's standpoint. Avoid becoming defensive; don't take the complaint personally or feel threatened. Confront the problem together. Ask the guest for suggestions on how to solve the problem. Then work together to find a mutually acceptable solution. An apology for having caused the guest any trouble is also in order.

Servers should try never to let a dissatisfied guest leave angry. If servers cannot solve a problem they should find someone who can. It is almost always worth it to give something to an unhappy guest, since that is an investment in future business. It might be a free item or free meal. If a guest ever becomes abu-

EXHIBIT 3.1 Most Frequent Complaints by Foodservice Patrons

1. Inconsistent and untimely flow of service
2. Poor food quality and improper serving temperature
3. Poor sanitation practices and poorly kept facilities
4. Staff not friendly and lacks product knowledge
5. Too long of a wait to be seated, served, and presented with the check
6. Cleanliness of service ware, utensils, and equipment not acceptable
7. Ambiance and decor lack character
8. Inconvenience of parking, accessibility, and location

sive or violent, it is best to call management, security, or the police and let them handle the situation. At all costs, try to avoid a physical confrontation. It is important to log all complaints in order to determine problem areas and avoid similar complaints in the future. You may have customers that just complain to get freebees, in this case, you can either accept it and count it as creating good will or refuse to do anything about it. The first is recommended.

Why Guests Complain

The reasons guests complain are varied. All complaints are legitimate to the guest making them, and can actually be beneficial to managers. Complaints can be opportunities to help improve one's business.

Some complaints are simply the consequence of a guest's disposition at that time. It could be a moment of swinging moods or an evidence of poor health. Some guests complain because of unrelated stresses. If work has gone poorly that day or the guest is upset over an argument, a guest is more likely to be impatient with small errors.

Management should see to it that servers are trained to handle complaints.

Personal issues may cause a guest to complain to a server who has done nothing wrong. Servers need to know how emotions are closely mixed with our food. As babies, we cry when we're hungry, in pain, want attention or in other ways are distressed. Given a bottle, we quiet down and take our comfort out by eating. This happens so often that food becomes ingrained in our beings as an emotional reliever. After babyhood, we still have this psychological tie-up. Thus, when there is a complaint from a guest about the food or something else and good investigation can find nothing wrong, the server should remember that this tie-up exists. Soothe as best you can the feelings of the guest. Often, the server can take the food back to the kitchen and the kitchen can change it a bit, maybe just reheat and replate it or merely add a different piece of parsley and send it back. Always remember in difficult situations to be patient, tolerant, and professional.

Signs of an Unhappy Guest

A server should watch the guests at her or his tables, noting their moods. If a guest looks irritated or unhappy or avoids eye contact, ask, "Is everything all right?" If a guest sits, not eating or drinking after a food or a beverage has been served, ask the same question. If the answer is a half-hearted OK, one might have a problem guest. If a guest asks for a menu and then checks the items served and sits, not eating and drinking, or sits looking around the room, also not eating or drinking, ask, "Is anything wrong?" A clear sign of a dissatisfied guest is if nothing is eaten or drunk or very little is, and they suddenly ask for the check.

Avoiding Complaints

Many complaints can be avoided by adhering to service standards. Servers must take orders carefully and communicate thoroughly with guests. For instance, if a preparation time is going to be quite long, the guest should be told. Remember that guests want explanations, not excuses.

At the start of the meal, some guests show they might be difficult. Astute servers can sense this and take special care to satisfy them. The server who gets the least complaints is the one who understands people and knows what pleases them.

If a complaint is caused by some action of the kitchen, the server should bring it to the kitchen's attention. A server should try to correct mistakes before they return to the guest's table.

FIGURE 3.1

The server should take any complaints about the action in the kitchen directly to the kitchen staff. Courtesy Action Systems, Inc.

Keeping Managers Informed

Management should be informed of all complaints even when they have been handled. A manager should be called in if a difficult one arises. Corrective action should be taken to prevent a reoccurrence. For example, if there are too many complaints of cold food, management must investigate, and any cause should be corrected. For instance, if the food is placed by the cooks on a countertop for pickup and stands there getting cold, management should find out why food is standing so long. If it's not the servers' fault, then management should purchase a **heat lamp** especially designed to be placed on the counter where food sits under it keeping warm.

A manager's apology for an error is always good policy and welcome. A server should avoid blaming another member of the staff. Managers should follow up on a critical problem so it does not reoccur.

Serving Guests with Special Needs

Servers must know how to serve guests with special needs.

People with Disabilities

A **guest with a physical disability** might use crutches, a cane, a walker, or a wheelchair. Often, members of the party will help the guest to his or her seat; the server may only need to take the walking aid and put it away in a safe place.

Sometimes it is useful for someone to greet guests at the door and bring them to the place where they will be greeted by the person who will see to their seating.

Crutches and canes should be stored near the guest's table, but not in the way, or directly under the table. Armchairs are helpful because people can put their hands down on the arms and lower and raise themselves. When the customer is ready to leave, the walking aids should be brought quickly to the table.

Guests in wheelchairs may only require that the server remove the chair at the table so the guest can pull up. If the guest requests it, the server can help him or her move from the wheelchair to the chair, and store the wheelchair out of the way. Some guest like to remain in their wheelchair. If so, remove the regular chair so the guest may take the space.

Those with a disability of the arms or hands may have special needs. Customers with some illnesses may be able to eat only with a spoon. Stemmed glassware is easily knocked over, so tumblers are preferred. Straws may be helpful. Serve hot beverages in a deep cup so any shaking of the cup does not spill the hot liquid. Some guests might carry eating utensils that are specially designed for them.

Communication Issues

Servers should pay careful attention to a guest who communicates unconventionally, or who nods or points to the item to indicate understanding. A small pad will be helpful for writing questions ordered, such as, "How would you like your steak cooked?" Another member of the party might give the order and avoid the need for writing out the question.

Some people simply have difficulty hearing. If the server faces the guest and speaks clearly, the guest often has no difficulty understanding. Again, the server may confirm the order by pointing to the menu, or by repeating the order back to the guest.

Guests with speech impairments might be difficult to understand. If a server has difficulty, he or she can ask, "Did I understand you to say . . . ?" to be sure. Ask questions courteously in order not to cause the guest any embarrassment.

Guests who speak little or no English may have difficulty getting servers to understand them. Someone on the staff might help. Body language, pointing, and nodding can help. Have the guest point to items on the menu, if some ability to read English rather than speak it is shown.

People with Vision Impairment

When serving a sight-impaired guest, be sure to greet him or her each time you approach the table, so your presence is known. Offer to lead a sight-impaired guest to the table. It is often helpful if the server offers an arm, but take your cue from the guest. At the table, pull out the chair and put the hand of the person on the back of the chair. This locates the person. If necessary, some guidance may have to be provided.

FIGURE 3.2
Guests with special needs may require extra attention.

Guide dogs are allowed in food services, and will usually follow a server to the table, guiding its owner. The seating is then the same, with the server now holding the dog's leash, while the guest sits. Lead the dog to a place near the table where it is out of the way, but can see its owner, or let it lay at its owner's feet. Do not pet, feed, or disturb the dog. When the guest has finished the meal, help the customer up, and let the dog lead its owner out.

When presenting the check to a sight-impaired guest, clearly state the total, then go over the individual prices. If using a credit card, you might want to restate the amount before the person signs the slip. The server may have to help the customer find the proper line to sign. If the guest pays with cash, state the bill denominations and coins of all change.

People with Special Dietary Needs

More and more guests dining out want foods that meet special **dietary needs.** This is partly because our population is aging, and as age increases special dietary needs increase, and partly because customers are becoming very health-conscious. Many food services are able to provide low-salt, low-carb, low- or high-fiber, low-cholesterol, low- or high-calorie, low- or high-protein foods. It is beyond the responsibility and mission of most food services to enforce healthful eating, but most make a variety of foods available to satisfy simple needs and tastes.

Servers should know enough about each of these dietary needs to assist guests in making selections and act to meet those needs. Guests will appreciate the assistance.

A good working relationship with the chef and cooks is essential in every application of service, particularly when dealing with guests who have special dietary needs. Special requests are more and more common. Many people ask for ingredients to be omitted, such as salt, oil, butter, or cream.

Managers should reward the servers who help their guests and ensure their repeat business.

CHAPTER SUMMARY

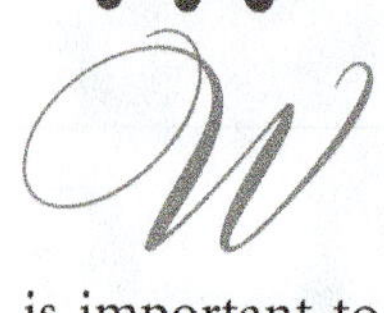

When dealing with guest complaints, the server should make sure to offer a sincere apology for any inconvenience. It is important to show sincerity, concern, and a strong desire to amend the situation.

Complaints can be helpful to the operation and the server. Accepting a complaint from a guest allows one to correct the problem and avoid similar situations in the future. The rule in receiving a complaint is to listen, accept responsibility, and resolve. If servers cannot solve the problem, they should find someone who can. By adhering to service standards, many complaints can be avoided.

Servers must know how to serve guests with special needs. Guests using crutches, canes, and walkers may have them stored near the guests table, but not in guests' way. Guests in wheelchairs may need the server to remove the chair for them, or help move them from the wheelchair to the chair at the table.

Special utensils may be needed by some guests with a disability of the hand or arm. Some may only be able to use a spoon, or may need to drink from a tumbler with a straw to avoid spillage.

When serving guests with hearing and speech impairments, servers should speak clearly, use a notepad, or point to items on the menu to clarify what the guest wants. If the server is unclear as to what the guest wants, he should ask again.

Vision-impaired guests may need to be assisted to the table. A server may need to describe the menu to the guest, explaining each item in detail. Guide dogs should rest near the table where they can see the owner. When presenting the check to a sight-impaired guest, the server should clearly state the total and individual prices.

Servers should be aware of various dietary needs, and be able to assist guests in selecting items to meet those needs. Having a good relationship with the chef and cooks in your operation will help procure the desired meal from the kitchen.

RELATED INTERNET SITES

Service and Guest Disabilities Council for Disability Rights
www.disabilityrights.org/fact_hotels

KEY TERMS

heat lamp
guest with a physical disability
guide dog
dietary needs

CHAPTER REVIEW

1. Describe how you would serve a guest who is deaf or hard of hearing.
2. A guest in your operation has a complaint about the slow service and poor food. How do you handle the situation?
3. Describe how you would serve a guest who is blind.
4. When receiving and accepting a complaint, what three steps should you be sure to follow?
5. A family of four has just been seated at a table in your section. The parents look tired, and the two children are crying. Should you be aware of outside circumstances that may cause the guests to complain? How can you avoid a complaint in this case?
6. Where should you place a guest's crutch, cane, or walker?
7. How do you seat a guest in a wheelchair?
8. List some ways to communicate with guests who have a speech impairment, or speak and understand little or no English.
9. How do complaints improve business?
10. You should never give anything free to an angry customer—it only encourages the person to continue behaving that way. True or false?

CASE STUDY

Establishing Rules

A new restaurant is to be opened, There is a need to establish a set of rules to guide servers in handling checks. Set up the rules you would like to see established for this restaurant. Include presenting the final bill and the handling of payment by cash and by credit card. Bring your list to class, and together as a group, decide on the batch of final rules you would want for this operation.

This operation is also going to need a set of rules to guide servers on personal hygiene, dress, and manner of dealing with guests. Include in your list rules for handling guest complaints. Bring your list to class and also together decide on the final list of rules in this area.

Set up a list of rules for what servers should do in handling emergencies such as fires, accidents at the table, guest illness, and so on. How should servers handle disabled guests such as blind guests with guide dogs, deaf people, or those who have trouble walking. Include how servers should handle alcoholic problems, and theft or other criminal actions. Bring the list to class and similarly compile a final set of rules.

Service Mise en Place

Outline

I. Introduction
II. Getting Ready
    A. First Impressions
        1. Lighting
        2. General Security
    B. The Dining Area
        1. Environmental Control
        2. Station Mise en Place
        3. Table Mise en Place
    C. Guest Checks
        1. Employee Theft
        2. Guest Fraud
    D. Getting Servers Ready
        1. Scheduling
        2. The Preshift Meeting
III. Clearing and Resetting Tables
IV. Ending the Meal
V. Quick-service Mise en Place
VI. The Cash Bank
VII. Chapter Summary
VIII. Chapter Review
IX. Case Study

Learning Objectives

After reading this chapter, you should be able to:

- Oversee the maintenance of well-stocked and organized service stations.
- Explain the importance of mise en place for servers.

INTRODUCTION

Many preparation steps must be completed before guests arrive to eat in a dining area. Not only must the place be clean and in good order, but equipment, food, beverages, dishes, and a host of other things must be ready for use. Without such prework, service becomes chaotic and disorganized. Not only are guests dissatisfied, but servers are frustrated and tired. Preparing well for the busy period of service that follows can pay off in making the work much more efficient, easier, satisfying, and financially rewarding.

The prework that takes place both in the kitchen and guest areas before service begins is often called **mise en place,** which means in French, "put into place." It means getting everything ready for serving guests, but it also means keeping things in good order as one works. In food production it means the same thing—have everything ready, and work to continuously keep things in order. Another commonly used term is *side work.*

Have a place for everything, clean up as you work, and think ahead. In both the service and production areas, good mise en place often denotes an effective worker—and a much happier one. Good mise en place makes work fun; the lack of it makes it a drudgery. Most importantly, good mise en place makes for satisfied guests.

Mise en place work is usually divided into three parts:

1. Getting ready
2. Sidework during service
3. Ending the meal

The first part consists of all work done before guests arrive, such as setting up the station, preparing ahead to meet guest needs, and arranging for service in an orderly fashion.

One group of guests may leave a table full of soiled dishes and linens, dirty ash trays, crumbs, and other things that need attention in order for a fresh, neat, correctly set table to be ready for the next group of guests. This is the second part, the first being greeting guests and seating them correctly. The last part is the work required in the departure of guests, closing the station and leaving it in good shape for the next guests. No one part is more important than the other. Each must be done correctly to achieve totally satisfactory mise en place.

Mise en place work differs for different serving situations. For breakfast, the table setting is different: perhaps a superfine sugar is put into sugar bowls rather than regular, table sugar. The condiments required might be minimal compared to those set out for lunch.

As an example of how a different situation can influence the mise en place, the type needed for counter service differs from that of a regular meal, and the mise en place needed for a buffet dinner is quite different from that of a banquet. In general, banquet service will not require clearing and resetting the table because there is just one setting. If another banquet, party, or meal is to follow one banquet, the mise en place is different and the work to prepare starts all over. In quick-service restaurants, employees and managers must be diligent in their efforts to keep guest areas clean, since guest turnover is quite fast.

Getting Ready

First Impressions

Upon arriving at a food service establishment, guests see the outside and the surrounding grounds first. This must be inviting and give a favorable impression of what is to follow. **Grimod de la Reynière,** editor of the first gourmet magazine, said, "The soup to a meal should be like a porch to a house or the overture of an opera; it should be an inviting prelude to what is to follow." The same could be said of the entrance and grounds of an operation, no matter whether it is a drive-in, restaurant, or a boat waiting at a dock to take an evening party sailing in the bay while serving dinner.

The grounds should be clean and in good order. There should be good lighting and good security. Signs should be well lighted and visible from some distance so guests can be prepared to stop. Landscaping should be attractive, well groomed, and appear fresh. Walkways should be uncluttered and in good condition, and guests should not have to walk too far to enter. If valet parking is available, it should be prompt and courteous. Valet employees should be neatly groomed and uniformed, and should assist guests exiting their vehicles.

Lighting

Good outdoor lighting is essential. Employees and customers need to see well in parking lots, on walks leading to the establishment's front and back doors, in stairways, near entrances, and in vestibules. These areas should be lit well enough to read a newspaper. Wire cages should protect exterior lights. A backup power supply is desirable.

Lighting serves as a deterrent to crime. Criminals do not usually want to be seen or identified, so they avoid a well-lit establishment in favor of one that has dimmer lighting. Lighting in hallways, storage rooms, and basements should also be bright enough so that anyone hiding in them can be clearly seen. Lighting should be sufficient such that no shadows are cast.

When patrons enter a foodservice operation, they will form an overall first impression. Lighting is a key element in that impression, along with color, sound, and décor. Quick-service facilities and employee cafeterias strive for a feeling of brightness and movement because they are busy places where people do not linger. Such a feeling helps contribute to higher guest turnover. In fine-dining restaurants, con-

FIGURE 4.1

The exterior of an establishment is the first impression for guests. The exterior should always be kept clean and inviting. Courtesy Corbis Digital Stock

versely, people come to dine slowly and relax. These operations need less light and should be quieter and slower-paced.

The effective use of indirect lighting can add depth to a dining room. Shadowing helps give a pattern and breaks up the lighting. The light, however, should not be so dark as to make it difficult for people to see as they dine or move about. Sufficient lighting should be provided at tables so guests can easily read menus. In addition, because of the danger of accidents, candles or burning lamps for table light or decoration must be properly protected and used with care. All table lighting should function properly.

In the kitchen, there should be enough light so there are no shadows on work places. Heatproof lighting is desirable in cooking areas. Fluorescent lighting provides the most uniform and lowest energy cost.

Lighting fixtures throughout the foodservice establishment should be cleanable. Light bulbs should be replaced on a regular schedule, while they are still giving adequate light. Lights are designed to burn efficiently for a specific number of hours. Studies have shown that changing all lights at this run-out time saves money over waiting to replace lights singly as they burn out.

General Security

Decor is an important part of making a good first impression. Many operations decorate entrances with pictures, plants, or art. The decor should not interfere with safe passage through the establishment. In full-service restaurants, the host stand should be neat, uncluttered, and out of guests' and employees' walkways. First impressions on entering the dining room should be pleasing and should give guests a welcome feeling of cleanliness and order. If the first impression is negative, it could linger on through the entire meal, and might affect an otherwise perfect experience.

Almost all operations will have a telephone available where guests can make calls. The area should be neat and clean, and phone books should be available. Check frequently to see that these books are in order and do not have missing pages. Also, make sure telephone areas are well lit.

All doors and windows should be clean and bright. The vestibule or hall should be well lit and attractive. If there is a coatroom or area for accommodating guests' outerwear or other items, it should be near the entrance. A sign should indicate that guests are responsible for the items left, but employees should be observant of any breach of security. Signs are no guarantee that the facility will not be held liable for theft or vandalism.

Foodservice operations are, unfortunately, not immune from criminal activity, regardless of how elite the establishment is. Pickpockets love busy places, and automobile vandalism or theft is a possibility. When a criminal activity is suspected, management should call the police and then investigate, detaining anyone suspected of committing the crime until the police arrive. Remove the disturbance to the manager's office and try to maintain a calm atmosphere in the establishment. Your establishment's attorney should provide employees with guidelines on dealing with suspected criminal activity.

The Dining Area

All guest areas should be neat, orderly, and inviting. Dirt or spills will spoil an otherwise perfect impression and negatively color a guest's feelings about your establishment. A dirty spot on the entryway carpet may spoil the cordial welcome the facility is trying so hard to create. That's why it's important for all employees to see that the areas for which they are responsible contribute to an overall favorable impression.

The decor of the dining area should be fitting to the ambiance and clientele of the operation. Simplicity at times can be better than too much lavish decor. In full-service operations, what guests should see is a neat, orderly grouping of tables, set with the proper flatware, decor, and glassware. (Some authorities recommend that glassware be turned upside down, while others object to this practice, saying that inverted glassware indicates the table is not ready for guests.) Candles and table decorations should blend in with or complement the decor.

Quick-service operations should have only clean, neat tables and booths. There should be a convenient place for depositing waste and trays.

Whatever the case, the arrangement of tables should be neat, not too close together, and clean. Chairs should be squarely set slightly under the table.

Before a meal begins, make an inspection to see that the dining area is ready for guests, and make any necessary corrections. Ask yourself, "Would I want to eat here?" Your servers might also ask themselves this same question. **Exhibit 4.1** is a checklist used in one operation to see if a dining room is in order.

EXHIBIT 4.1 Sample Dining Room Checklist

Answering yes to each item indicates excellent mise en place.

DINING ROOM CHECKLIST

Inspected by ____________________

Time and Date ____________________

Area ____________________

Temperature __________ Humidity __________ Air Quality __________

Yes	No		Yes	No	
❑	❑	All lights turned on?	❑	❑	Traystands stored in proper locations?
❑	❑	Fixtures clean?	❑	❑	Silencer on properly?
❑	❑	Windows clean?	❑	❑	Table linen clean and pressed?
❑	❑	Draperies properly arranged?	❑	❑	Table cloth arranged properly?
❑	❑	Furniture and decorative items clean and dusted?	❑	❑	Napkins properly folded and placed?
❑	❑	Stations inspected?	❑	❑	Table lamps clean and in good working order?
❑	❑	Tables level?	❑	❑	Flower vases clean and filled with fresh flowers?
❑	❑	Tables arranged properly?	❑	❑	Flowers nicely arranged?
❑	❑	Aisle space adequate?	❑	❑	Covers set properly for meal or occasion?
❑	❑	Chairs in good condition?*	❑	❑	Dishes clean?**
❑	❑	Clean?	❑	❑	Flatware clean?
❑	❑	Chairs arranged properly at the tables?	❑	❑	Glassware clean?
❑	❑	Carpeting clean?			
❑	❑	Floors clean?			
❑	❑	Floors polished?			

**Check legs for splinters; check seat covers for rips and stains.*

***Check each setting for proper flatware and placement; check also for proper placement of dishes, etc.*

Environmental Control

The lighting, air conditioning, heating, and ventilation should be comfortable. Check for drafts. Be careful to eliminate any unpleasant cooking odors in the dining area.

Managers should know the basics of mechanical adjustments so they can help maintain a comfortable environment and correct problems. A manager should always be notified of heat and humidity problems.

Humidity can cause guest and employee discomfort. When the air is too moist, it does not evaporate well from the body, and this causes body heat to build up. The air in a dining room should have around 50 percent relative humidity. To check whether the dining area is too humid, just fill a glass with ice and water and set it down on a dish. If beads of moisture quickly build up on the surface, the room is too humid. Treat air to remove excess moisture as well as cool it down.

Servers should check to see that tables near cool air outlets are not blasted with cold air. Nobody wants to sit in a cold draft. Direct sunlight coming through a window quickly builds up heat. Proper window coverings can control the sun's rays. A desirable temperature in the dining area is around 70°F (21.1°C). If the temperature is hot and dry outside, the temperature can be somewhat higher and guests will still feel comfortable. Dry air helps people feel cool even though the temperature may be higher than 70°F.

Those in charge of dining areas should understand that guests themselves give off heat and moisture from their bodies and from the air they are exhaling. It is said that every adult gives off as much heat as a 100-watt light bulb. Thus, in a well-lighted dining room filled with guests and servers, heat can build up considerably and must be monitored to ensure everyone's comfort.

Station Mise en Place

Dishes, flatware, glassware, and linens are usually kept in a pantry or storeroom, and servers may have to retrieve these items before each meal. A steward is usually in charge of these items in large, formal operations. In some operations, these items might be located in the service station or on the service table near the dining area. This helps to speed service. Normally, servers get their own supplies for their stations, but buspersons may help them. Often, the server sets up the original stock and then, during the busy time of service, buspersons maintain supply levels and bring items to the station sidestand, or to the tables themselves.

FIGURE 4.2
Servers should have access to dishes, flatware, glasses, and linen to reset tables quickly between guests.

Besides dishes and linens, salt and pepper shakers, condiments like mustard, ketchup, steak sauce, oil and vinegar in cruets, filled sugar bowls, and other items must be there, if their need is expected. All bottle necks on condiments, sugar bowls, salt and pepper shakers, and other items should be clean. Special attention should be paid to necks, mouths, and caps of bottles. **Exhibit 4.2** lists a suggested inventory of items for a sidestand.

Clean dishes can be brought to the sidestands on trays. The bottom of the tray should be lined with a clean cloth so dishes do not slide as they are carried. Be careful not to stack items too high, they may tip over and cause an accident or at least cause a server to lose time, and breakage can be costly. Cups and glasses can be brought to a service station in the racks they were washed in or on a tray. They should be placed inverted on the trays or in the racks.

Dishes, glassware, and other items that show cracks or chipping should be discarded, as should bent or otherwise damaged flatware.

Dishware, flatware, and glassware should be clean and sparkling. However, just because something looks clean and sparkling does not mean it is sanitary. Bacteria can be present without being seen. That is why these items are given a last minute or two of sanitizing while in the washing machines. In sink washing, the final rinse is extremely hot and usually includes a sanitizing agent. Wiping ware dry is not recommended. Health authorities frown on it because it can pass contamination from one utensil to another.

High temperatures and dishwashers using sanitizing compounds aids in providing a high volume of guest-safe dishware. The following steps are recommended:

1. Check cleanliness of dish machines and see sanitizing equipment are operating correctly and are filled with an adequate supply of sanitizing medium. Wash and rinse tanks should start with fresh, clear, hot water.
2. Scrape, flush, or soak items before washing. Pre-soak with dried or cooked food on items.
3. Use racks constructed so all item surfaces are exposed for cleaning. Correctly load machines. Do not overload. Washing and sanitizing efficiency is increased. Second pass throughs are avoided.
4. Check temperatures frequently.
5. Check items as racks are unloaded. Run items needing it through again. Proper equipment and proper work methods insure one-pass dish washing.
6. Air dry everything; do not use towels.
7. Keep machines in good repair.

Hard water can cause spotting on items, and over time, a mineral buildup can occur. This spotting and mineral buildup can be avoided by using the proper detergents and other washing compounds, usually a compound associated with phosphoric acid. If spotting cannot be avoided, wiping might be necessary to remove the spots. Mineral buildup is not as easily taken care of. Soft abrasive material is needed to remove the built-up soil, and it can be very time consuming. For these reasons, preventing the buildup in the first place is the best practice. Detergent companies often can be of help in solving these problems.

Sterling silver flatware and other items should be well polished. If the operation has a silver polisher, these items will come from the dishwashing section in satisfactory condition. If not, servers themselves may have to polish or touch up some pieces. Soiled flatware should be separated at the service station. Some operations allow flatware to be kept in the containers in which they are washed with the handles up. This is sanitary, but may not be very presentable.

FIGURE 4.3

Servers should keep their workstations clean and organized, the better to respond to guests' needs.

In some types of service, additional equipment will be brought to the table. For French service, a *guéridon* with all the necessary items will be brought tableside so the *chef de rang* may prepare the items. Service carts or wagons (*voitures*) may be brought to the table to display items for guest selection, or a carving station or other mobile equipment may be brought to the table. They should be so clean they shine, and metal parts should be polished. Glass or plastic covers should be clear. All equipment—tray stands, high chairs, and other equipment that may be needed from time to time—should be clean, neat, and in good working order.

Servers must often operate hot and cold food and liquid holding equipment. Follow these procedures to ensure the equipment is functioning properly:

1. Hot food should be kept at temperatures above 140°F or higher. Holding equipment includes steam tables, double boilers, bain maries, heated cabinets, and chafing dishes.
2. *Never* use hot-holding equipment to cook or warm items—use only to keep foods hot.
3. Measure equipment temperatures every two hours and enter temperatures on a log.

Cold and raw foods and cold liquids must also be kept at proper temperatures to avoid contamination and retain textures and flavor. Servers should do the following:

1. Hold cold foods and liquids at about 40°F or lower.
2. Hold ready-to-eat cold foods in pans, plates, or other suitable containers, never directly on ice. Be sure ice surrounding food drains away as it melts. Sanitize drip pans after a day's use.
3. Measure temperatures every two hours and log them.

Before the station is ready, all food items needed by servers, such as butter, cream, coffee, ice, water, condiments, and garnishes should be on hand and properly stored.

EXHIBIT 4.2 Example of a Standard Inventory for a Station of 20 to 25 Covers

All quantities are recommendations. The items and quantities in individual operations will vary.

FLATWARE			
Dinner forks	30	Soup spoons	24
Dessert/Salad forks	30	Cocktail forks (dinner only)	12
Dinner knives	30	Grapefruit spoons (breakfast only)	12
Teaspoons	30	Fish knives (lunch and dinner only)	12
(Other flatware according to menu needs)			
DISHWARE			
Dinner plates	25	Salad plates	25
Bread plates	25	Cups and saucers	25
Soup bowls	15	Bouillon or soup cups	15
(Platters and other dishware according to menu needs)			
GLASSWARE			
Water glasses or goblets	30	White wine glasses	15
Red wine glasses	15		
(Other glassware as needed)			
CONDIMENTS			
Ketchup bottles	4	Dijon mustard	2
Worcestershire sauce	2	A-1 sauce	2
Tabasco sauce	2	Soy sauce (If needed)	2
(Chutney, oil and vinegar, grated cheese, etc. as needed)			
REFRIGERATOR			
Milk, regular	4	Milk 2%	4
Cream or substitutes, indiv.	36	Butter pats (144/box)	1 box
(Lemon, etc. as needed)			
LINENS			
Napkin, folded	30	Naperones (if used)	5
Tablecloths	5	Hand towels	5
Serviettes	5		

MISCELLANEOUS

Sugar substitute packets, salt substitutes, ice with tongs, syrups, jellies, jams, water pitchers, water, ash trays, pencils or pens, menus, peppermills, plate warmers, coffee makers, coffee packets, coffee maker cleaners, various trays, check trays, finger bowls, crumbers, tote boxes or buspans, etc., as needed.

Usually such storage is close to the station and other servers may use it so the servicing of this area in mise en place is a joint responsibility. Buspersons might do this.

Frequently servers operate the coffee equipment. Before service starts, there should be a check to see that coffee and other items needed are on hand. Usually a 12-cup filter unit will be used. Coffee should always be fresh and prepared properly. Once a batch is made, remove the grounds immediately. If the grounds are allowed to stand over the brew, bitter compounds can drip down from them and reduce brew quality. Coffee should be held no longer than 45 minutes.

Table Mise en Place

The first thing a server in a full-service restaurant should check before placing linens on tables is the tables themselves. Are they sturdy and not wobbly? Is gum deposited under the edges? Are they clean? Are they level? Are chairs clean and the seats free of crumbs? Is the area around the table clean and in good repair? If not, any undesirable condition should be corrected before the table is set.

Some operations cover their tables with a **silencer** and tablecloth. A silencer is a felt pad that quites the noise of the table service and absorbs spills. The server brings the silencer and tablecloth to the table and places them on a tray stand or another table. If items are on the table, these are removed and placed on a tray stand, cart, or service station counter. The silencer is then placed on the table first; it might just fit the shape of the tabletop or it might hang down not more than eight inches over the side. To lay the silencer, pick it up at the centerfold at both ends and, with arms extended, lay it so the centerfold is in the center of the table and the silencer covers half the table on the side opposite you. Unfold the other half and place it so the table side on your side is covered. Sometimes the silencer may be given a quarterfold rather than a halffold. The procedure in this case is much the same as with the centerfold in the center, but the two top folds must be lifted up with the fingers and placed so as to cover the half of the table opposite the server. Now the underfolds must be carefully pulled from under and moved to the table edge where the server stands.

The method for laying the cloth on the silencer is similar to that used for laying the silencer. Smooth the silencer out before laying the cloth. The tablecloth should hang down at least eight, but no more than inches over the table sides, with hangover equal on all sides. Some may wish to have the hangover so the tablecloth edge almost touches the chair seat. Smooth the tablecloth out and step back and check to see that the cloth is centered. If not, adjust the cloth. Examine all table coverings for holes, cigarette burns, and stains and replace them if necessary. A naperone, also called a laycloth, may be laid on the table in a similar manner to that used to lay the silencer and tablecloth.

The placement of napkins can vary. Some operations place them in a stemmed wine glass, although some health inspectors have raised concern because lint particles can remain in the stemware. Others place them folded in the center on the service plate, if used, or on the tablecloth between knives and spoons and forks, if no service plate is used. Place plainly folded napkins, either paper or cloth, with one corner facing the guest, so when the guest picks up the napkin, it can easily be unfolded with one hand. Often in fine-dining establishments servers pick up the napkin and either hand it to the guest or place it across the guest's lap. Napkin folding has reached a high art of skill and many operations have intricately folded napkins that add to the table decor.

If the table is to be set with place mats or have no covering at all, wash the table top with a mild detergent in warm water and wipe dry. The tabletop should shine and have no sticky or soiled areas. Next, place the place mats squarely about two inches from the table edge. If these are the type that are reused, see that they are clean and free from grime or stickiness.

Items that are to go at the table center, such as flowers, candles, and salt and pepper shakers should now be placed on the table, as well as ashtrays, matches, table tents, and any other items. Then, proceed to set the table with the necessary covers. It is accepted practice to place flatware in the following manner: forks on the left, and knives (with blades facing inside) and spoons on the right. A dessert spoon or fork may be placed in the cover center above the napkin. Normally, flatware is set in place with the first piece to be used on the outer edge, and then as the meal and need for items proceed, the guest uses items from the outer edge inward.

The bottom ends of knives and forks should be placed at least one inch away from the edge of the table. When placing silverware on the tables, use a plate lined with a napkin to hold the ware. In casual dining, servers may use a cocktail tray, a plate, or simply a clean service towel as an underliner; servers should avoid walking around the dining room holding flatware with bare hands. Touching flatware by any part other than the handles should be absolutely avoided.

If the table setup requires it, the bread and butter plate is placed next to, but slightly above, the salad or appetizer fork. A butter knife is placed across the plate horizontally. Some operations prefer placing the butter knife vertically, so the tip of the knife points in the same direction as the dinner knife. In this case, the blades of both knives should face left.

The water glass is placed near the tip of the dinner knife. The wine glass (if required) is placed to the right of the water glass. If the operation offers two different shaped wine glasses (one for white wine, and one for red wine) the smaller of the two (usually the white wine glass) is placed slightly above the water glass, and the taller glass is placed slightly above and between the two. This might seem unimportant at first, but it will make a difference when serving wine, as it is difficult to pour any liquid into a glass when a taller glass is in the way.

FIGURE 4.4

It is important to have all tables set properly before the dining operation is open to guests.

If additional glasses are used, such as those for sherry, champagne, and brandy, they should be set toward the center of the table, so that they will not cramp the space available between guests. This should be kept in mind when setting a table for four or more.

Whether in casual or fine-dining operations, symmetry is a must. An effective method to ensure symmetry in flatware placement is to draw an imaginary line between the tip of the dinner knife and the dinner fork on the opposite side of the table. If the line is straight, the setup is symmetrical. This can be applied to any table shape and size, with the only exception being a round table where an odd number of covers are placed.

Centerpieces and condiments should not be placed in a manner inconvenient to guests; for example, flower vases should not keep guests from seeing others sitting on the opposite side of the table. Additional items that might be requested during the meal should be placed conveniently to the right side of the guest.

In some operations, such as a cafe, the table setup is simpler. Often there is no linen on the table and paper napkins may be used. A sugar bowl, salt and pepper shakers, ashtray, and matches (if there is a smoking section), and perhaps a bud vase or table decoration and some condiments are placed in the center. If the table is set against the wall, these items are placed on the edge of the table near the wall. Paper napkins may be in a container. Sometimes the table is set after the orders are placed, the server then bringing the correct items such as paper placemats, or flatware settings wrapped in paper napkins.

Often in coffee shops, cafes, or other busy operations, coffee cups might be on the table for breakfast. The flatware is usually limited to a knife, fork, and teaspoon. Other flatware will be brought as needed, such as a steak knife, soup spoon, iced tea spoon, or dessert spoon or fork. Tables are often not covered with linen for breakfast or lunch—placemats might be used—while linen might or might not be used for dinner.

In most quick-service and self-service operations, the table is bare and guests bring to the table the items they need. However, this may vary; some may still set tables with a limited amount of items.

Exhibit 4.3 contains a list of some basic mise en place instructions that will assist in any service style setup. **Exhibit 4.4** is a reminder list of common side work.

Guest Checks

Order or guest check systems vary from operation to operation. Some restaurants still issue servers a supply of guest checks. With computerized POS (point-of-sale) systems, which are used by many restaurants today, this usually is unnecessary. Instead, checks, properly coded as to server, station, and table, are given to the server at the time guests are brought to the table, or are produced once order information is entered into the computer.

EXHIBIT 4.3 A Partial List of Mise en Place Instructions

1. Be on shift and station on time.
2. Be sure to punch in on the time clock when you arrive.
3. Be prepared regarding personal appearance and uniform.
4. Check your station for adequate supplies of glassware, garnishes, straws, napkins, picks, etc.
5. See that all equipment is clean and ready to use.
6. Sign in for and receive change money.
7. Check station to make sure all areas are in proper condition (tables/chairs clean, arranged properly, floor clear, ashtrays on table, etc.)
8. If you are relieving another server, check with her/him for any instructions as to guests that might be in the station, pending orders.
9. The service area at the bar is to be kept clean at all times, free of litter, soiled towels, dirty ashtrays, and so on. All garnishes and supplies must be stored in an orderly fashion. Wipe the bar service area frequently to keep it clean. Keep garnish containers clean and filled with fresh items. It is advisable that different servers on each shift be made responsible for specific service area tasks; thus, one server can be made responsible for garnishes, another for picks, napkins, and so on.
10. Be sure you have a bottle and can opener at your station.
11. Be sure the ice bin area is kept clean. Use the scoop whenever filling glasses with ice. Every station should keep about a dozen glasses iced up.
12. Drink tower areas and sinks for drawing fresh water should be kept clean. Do not wash ashtrays under drink towers. This may create serious drainage problems.
13. Always use a large service tray in giving service. Prepare and stack no more than six trays. Arrange service items on trays on one side of cocktail tray. Each tray should contain:
    - ashtrays (if the restaurant has a smoking section)
    - tip trays
    - matchbooks
    - pens
    - bar towel
    - cocktail napkins
14. Check ashtrays frequently if the establishment has a smoking section. Remove as necessary, replacing with clean ones. Use the capping method by placing the clean ashtray on top of the dirty one; remove from table and place the dirty ashtray on side tray. Then put clean ashtray back on table.
15. The floor around your station must be kept free of litter (straws, napkins, empty sugar packets, etc.)
16. Unoccupied chairs should be pushed close to the tables to allow free movement between tables. This will also help to keep the station organized.
17. At the end of the shift and at closing, all tables and chairs should be returned to their proper places as per station chart. They should all be wiped clean with a bar towel.
18. At the end of a shift, all ashtrays should be washed and returned to their storage area in the station.
19. S-O-S (stay on station). Be ready to offer more service. Take that second order. Pick up soiled napkins, discarded stirrers, and so on. Put chairs back in place.
20. Circulate; don't congregate.
21. Utilize the round trip. Don't return to the station empty handed. Clear tables before going to the bar. With light loads from the bar, bring station supply replenishments.

EXHIBIT 4.4 Side Work Reminder

A. Condiments
  1. Fill partially full containers with other partially filled ones. Fill to half-inch from the top.
  2. If containers are to be refilled, wash and dry first and then refill.
  3. With a damp cloth, wipe tops of containers; clean caps and replace.
  4. Store perishable condiments in neat rows in the refrigerator.
  5. Day servers will empty and refill relishes.

B. Sugars, salts, and peppers
  1. Fill and wipe salt and pepper shakers; make sure holes are clear.
  2. Refill sugar trays. Be sure old packets are not allowed to rest on the bottom under the row of packets on the tray.
  3. Wipe holders.
  4. On Monday nights all containers and trays should be emptied and washed. Refill only after being sure the containers are dry.

C. Butter and cream
  1. Don't use broken or partially melted butter patties. Place unusable patties in a dish to be taken to kitchen for use in cooking.
  2. Keep butter containers iced and free from melted water.
  3. Return emptied butter containers to pantry for washing and refilling.
  4. Keep coolers clean at all times; wipe out as needed. On Sundays, empty and wash, refilling with butter and cream.
  5. Fill creamers to about half-inch from the top. Do not refill creamers; use only clean ones.
  6. At end of shift, empty filled creamers into cream containers and send the emptied creamers to the dishwashing area.
  7. Keep at least two containers of cream and two filled butter containers in the cooler; more may be needed at busy times. Pick these up at the pantry.

D. Dining room buffet and water cooler
  1. Have placemats, napkins, salt packets, sugar packets, and takeout containers on hand.
  2. Have clean water pitchers on hand.
  3. Stock all these items on the shelves under the buffet.
  4. Stock tip trays and ashtrays on top of buffet. Keep the remainder of this area free for use during service.
  5. Stock glasses under water cooler.
  6. Remove daily the tray under spigot and wash drain area. Once a week use drain cleaning solution. Wipe down entire cooler daily.

E. Coffee urn area
  1. Clean iced tea dispenser and tray daily.
  2. Empty ice cubes for closing work; return lemon slices to kitchen pantry.
  3. Stock ramekins, iced tea glasses and spoons, teapots, children's cups, cups, and saucers on undershelves of coffee urn stand.
  4. Stock coffee, tea bags, instant tea bags, and coffee filters on undershelf of urn stand.
  5. Wipe entire area down daily.
  6. Clean coffee urn thoroughly at the end of each emptying. On closing, leave urn filled with water that has had a fourth cup of soda added. Mix well after adding. Rinse urn thoroughly in the morning after emptying and before using. On Saturdays add one bag of cleaning solution instead of soda and mix well. (Store this cleaning solution in box on bottom shelf.)
  7. Send bar trays through dishwasher once a week.

F. Bread warmer, boards, and knives
   1. Wipe bread boards and bread knives with damp cloth after each use.
   2. On closing, empty all trays of bread supplies and take to bake shop. Clean drawers. Be sure they are free of crumbs.
   3. Wipe entire outside of warmer daily.
   4. Have a good supply of napkins.
   5. Clean bread baskets once a week.

G. Salad dressings
   1. On closing, empty dressing jars into their storage containers and place the storage containers back into the refrigerator.
   2. Stack the empty containers on a tray with ladles in between and take to dishwashing section.
   3. Wipe racks with a damp cloth daily. Empty and clean refrigerator once a week.
   4. Store condiments on tabletop. Clean tops and store perishable ones in refrigerator on closing.
   5. Store French and Italian dressings on shelf under shrimp cocktail containers.

Where paper checks are used, servers are usually given a set quantity of numbered checks. Servers sign for them and are responsible for seeing that guest checks are properly used and that they get to cashiers with payment. In some operations, checks are issued to servers at the time guests are seated. These checks are kept at the cashier stand. Servers sign for the checks given to them at the start of the shift, usually next to the check number on a form commonly called the **traffic sheet.**

Employee Theft

Employees may be able to steal money by manipulating guest checks. Managers should watch for a slower flow of checks, or missing checks from those given out to servers, fake checks among legitimate ones, or checks that are regularly turned in late by an employee.

Other ways of stealing follow:

- Destroying checks, not ringing up the sale and keeping the money.
- *Bunching checks* by ringing up only one of several checks with identical amounts, collecting all the money, and keeping all the money except that for the rung-up check.
- Giving guests incorrect change.
- Raising prices on checks or charging for items not served.
- Stealing guest checks from the establishment's supply, and using these and never turning in the check and pocketing the money. False checks may also be brought in.
- Taking payment money left on another server's table.

One way to help protect against theft is to have the server present the check, receive cash or a credit card, and take this to a cashier. The cashier rings up the sale and gives the server any overpayment, which is returned along with the rung-up check to the guest. The guest signs the credit card receipt, giving it back to the server so the establishment has something to offer for its payment. The payer usually puts the tip amount on this signed receipt. If it's a cash deal, the guest paying the bill picks up the money returned, usually leaving the tip money on the tray for the server while keeping the rest.

An inspection should be made to see if all checks given to a server are returned either as used or unused checks. Some operations require servers to pay a certain amount for missing checks. POS systems as previously described alert management to problems. If guests are given checks to be taken to cashiers, servers are not considered responsible for missing checks. Cashiers are.

Some servers are not alone in finding creative ways to defraud restaurants. Some guests wait until there's a crowd at the cashier's station and then skip out on the check. It is the obligation for any employee to see that this does not happen.

Guest Fraud

Other forms of fraud involve lying about the bill. Diners may purposely fail to report when their bar bill is not carried over to their food order. They may erase or alter entries on checks. Careful register use, stapling of all related checks, and keeping control of the checks as long as possible can help avoid check trouble. Customers may sometimes claim they were given insufficient change. To avoid this, servers should leave the change on a plate and name the change amount as the money is set down. Cashiers should leave the payment on the cash register and then pay the guest the correct change, naming the amount given as payment is made.

Getting Servers Ready

Scheduling

In full-service restaurants, specific mise en place planning must be done by the host, hostess, maitre d', or other head of service. This includes forecasting server staffing needs. This is typically done by estimating the total number of covers, and then dividing this total by the number of covers one server can take care of during the meal period. Thus, if an operation estimates 120 guests for lunch, and the number of covers a server is expected to serve is 40, the number of servers scheduled is determined: $120 \div 40 = 3$. The forecast may be made for a week, and daily work assignments should be made so servers know when they are to be at work (see **Exhibit 10.4**). Servers should be informed of their schedules well in advance so they can plan their other priorities.

It is important to plan the number of servers needed and the station assignments on a practical and realistic basis. Each operation's staffing needs differ based on the size and type of the establishment. Only the number of servers needed should be scheduled, and no more. If more customers come than were expected, or a server misses work, a well-organized staff of servers can smooth over the unexpected and still have happy, satisfied guests. It is important in such cases that the service staff be well trained to ensure that things run smoothly.

Wisely, some operations maintain a list of *extra board* service employees, which are not included in the payroll on a steady basis but are called in case of emergency. Larger operations use an on-call system.

The person in charge of service will assign stations. Station assignments let the servers know which stations they will be working, so they can move ahead and complete the necessary mise en place. Often a chart is posted showing station locations and the name of the server(s) responsible for them. As the customer flow fluctuates, stations may be restructured or reassigned. This allows for servers to cover stations while others take a break or end their shifts.

The Preshift Meeting

Operations often have a short meeting of servers or all employees—sometimes called lineup, or preshift meeting—just before the meal period begins. Tables should be set, all other arrangements should be completed, and only last-minute things should be done. The meeting is usually conducted by the manager, host or hostess, or a supervisor. Often a quick inspection of grooming and dress occurs, but this is informal.

This preshift meeting is to go over the menu and any details of service that need to be brought to the attention of the service staff—whether any menu items are not available that day, the soups or fish of the day, and so on. Perhaps some important guests have made reservations or a special group is coming in which should be brought to the attention of servers in the station where they will be seated. Any special menu items should be covered with a description of how they are prepared and served. The correct items to use with various dishes and the manner of the service of the items, along with price and preparation time, should be included.

Each server should copy this information down since it may not appear on the menu and all information will have to be given verbally. Specials are usually items management wishes to promote, but others on the menu may be highlighted. Planned runout times may be given and suggested substitutes indicated. A bulletin board in the kitchen can give some of this information. A visit may be made to the kitchen where the chef may briefly add any details he or she wishes the server group to hear.

Sometimes the meeting covers complaints or compliments received. These can be helpful in indicating trouble spots and also indicating what guests like about the service or food. Since the time that follows can be stressful, the meeting should end on a positive note so servers are motivated to deliver great service.

Clearing and Resetting Tables

When a full-service dining room is full and other guests are waiting to be seated, tables cannot remain vacant long without angering customers. Clearing and setting tablecloths properly is important in maintaining a good image for the establishment. Thus, servers, buspersons, hosts/hostesses, and managers must work together to ensure that tables are cleared and reset in a swift, appealing manner.

Changing the linen is actually more difficult than the original set-up. One recommended method is as follows: Come to the table with a clean, folded cloth. Avoid holding it under the arm. Remove all items from the table to a tray, cart, or service station. Place the folded cloth on top of the soiled one. Next, pull the soiled cloth toward you so the hem of the soiled cloth is even with the far edge of the table. Unfold the fresh cloth by holding the center crease and the top hem between the index and middle finger. Then gently flip over the bottom section. Now hold the top edges of both cloths between the thumb and the index finger. By pulling on the opposite direction, you will pull the soiled cloth out while the fresh cloth remains in place. Step back to make sure the new cloth is centered and make adjustments as necessary. This method allows you to change cloths without exposing the naked table surface to patrons.

Another popular technique is to first place tableware items on the far side of the table. Then, with the forefingers of either hand, hold the two corners of both the soiled and the new cloth, pull up toward the center of the table, slightly raising both cloths' edges, and lower your fingers so as to fold the new cloth in half under the soiled one. Holding the corners of both cloths, fold them over again in the opposite direction. This should leave you with the soiled cloth folded in half under the clean one. Pull out the soiled cloth in one swift motion from the other side of the table. The tableware items are then returned to the center of the table.

FIGURE 4.5

In preparation for new guests, tables should be reset quickly and properly soon after guests leave.

Some other methods are faster and more suitable for a large-volume operation, but they involve uncovering the table. A common method is to fold back one edge of the soiled cloth and move the tableware items to that corner. The soiled cloth should be folded together to seal in the crumbs so they do not fall onto chairs or the floor. Place the clean cloth on the table, using the unfolding technique described earlier. Move the tableware items back to the center of the table and straighten the new cloth. With this and all methods, if the table is against the wall, pull it a few inches away from the wall for easier clearing.

When a bare table is used, wipe it clean with a cloth dipped in warm water with detergent, wring completely dry.

Reset the table with the proper dishes, flatware, and glassware, and any items that had temporarily been removed. Place napkins accordingly. Once the table is reset, it is ready for the next guests.

Ending the Meal

Mise en place is also required to prepare for guest departure. The bill should be totaled and ready to deliver. Some facilities deliver the bill by placing it face down on a small tray in front of the person paying, or in the middle of the table if the payer is not known.

If the payment is by credit card, provide a pen for writing down the required information. A rapid return with the change or the credit card and credit charge slip should be made. Any personal items that servers have taken care of during the meal, such as crutches, should be quickly brought. Help may be needed to put on coats. A sincere thanks from the server is appreciated. As the guests leave the dining area, the host, hostess, maitre d', or other individual at the entrance thanks them and invites them to return. In very informal situations, such as a quick-service operation, this may not occur. In those establishments, payment is made when the guest gets the food, and there is little formality needed on departure. Nevertheless, when guests are leaving, a warm thank-you makes all the difference.

The departure of guests should be given as much attention and care as the arrival. Cashiers, coatroom attendants, managers, doorpersons, valet parkers, and others should be friendly and helpful.

As a shift winds down, servers should complete their sidework. All soiled dishes should be taken to the dishwashing area and clean stock should be replenished. Soiled linen should be taken to its designated receptacle. Salt and pepper shakers should be filled, cleaned, and checked to see that holes are not plugged. Sugar bowls should be filled and cleaned. All other condiment containers should be filled and cleaned according to the operation's standards. Periodically, all salt and pepper shakers, sugar bowls, and other equipment should be emptied, washed thoroughly, dried, and then refilled. Check the sidestand to see that it is clean and the inventory is up to par. Old coffee should be emptied out and the containers cleaned. It might be necessary on a regular basis to put into the coffee machine a compound that frees it of lime buildup.

Health regulations in communities differ and must be followed. Perishable condiments should be placed in a refrigerator. Iced or refrigerated butter may be returned to refrigeration. Items not in sealed containers that have been put onto guest tables should never be reused for service. However, items such as jams or jellies in sealed containers are reusable. Be sure to know the expiration date for all perishable products and discard any items that have not been kept sanitary and at a safe temperature.

The closing work differs according to the operation and the meal. In most operations, closing includes setting up the dining and service areas for opening the next day. Each operation should establish complete checklists that outline what must be done at the end of each meal period to prepare for the next. For maximum efficiency, tasks may be assigned to individuals based on service area or shift length, for example.

The inspection checklist in **Exhibit 4.5** can be used at closing, as well as before guest arrival. Servers should leave their stations clean, stocked, and prepared for the next shift. The next server will appreciate help on getting a good start. Likewise, each server profits from the servers who previously worked the station.

Quick-service Mise en Place

Attention to detail is as important in quick-service operations as it is in full-service restaurants. The primary servers are the counter employees and employees out on the floor cleaning and stocking amenities. When taking the order, the server should repeat the order so the guest can hear and verify it before it is totaled up on the preset register. Payment is usually taken before the food is served.

When the order is assembled and put on a tray or in a bag, the guest typically takes it to a table or a service area for eating utensils, napkins, and condiments. In some operations, guests bus their own tables and dump their own trash. In others, employees perform this task.

In operations with drive-through pickup windows, the employee is responsible for helping guests select items, taking and placing orders, expediting, and giving a courteous, positive impression to guests.

To give such service requires precise and considerable mise en place. At peak service times, long lines challenge servers to work quickly and accurately. Without good mise en place, quick service is doomed. For service staff, this means that plenty of bags, wrappers or containers, condiments, coffee stirrers, cream and sugar, napkins, and straws must be on hand. Cash drawers should never be allowed to get low. Only the very last-minute details of assembly should be left. Because of the speed and efficiency expected by guests, mise en place is crucial.

The Cash Bank

In some operations, cocktail and other servers carry their own money or **cash bank.** This allows quick payment of a check and reduces the amount of work required of the staff. Procedures for some servers may have to make cash transactions from a cash fund. If used, one of the first steps of coming on shift is to obtain a cash bank to use in making change when collecting bills. The server should be sure to verify the correctness of the sum given.

Servers may carry the bank, but often the bank may be kept at a cash register in a drawer that only that server can operate. The register must be turned on and the server must enter his or her server number, perhaps a password or drawer number. The check for payment will be inserted into the machine, allowing it to read information on it, such as check number, table number, server number, items on the check with their price, and check total. The amount of payment is entered; the proper sum is put into the register and the drawer is closed, which closes the transaction. The check is filed in a special spot or may be kept by the server for closing out the cash bank at the end of the shift.

Sometimes cash is not given by the guest, but some other type of payment is used. A common way of paying in clubs and hotels is to charge to one's account. In clubs, no identification is usually needed. If one is needed, the club member usually makes identification by offering a club membership card. In hotels, the person charging the bill may show for identification a key or key card or some card issued at the time of registration. If there is any doubt as to identification, a call can be made to a front desk or other agency to verify a name, room number, and other pertinent information.

EXHIBIT 4.5 Sample Closing Checklist

Inspected by ____________________

Time and Date ____________________

Yes	No	
❑	❑	Lights off except designated ones?*
❑	❑	Air conditioning turned off?
❑	❑	Proper doors, windows, etc. locked?
❑	❑	Alarm set properly?
❑	❑	Tables cleared properly?
❑	❑	Chairs inverted on tabletops?
❑	❑	Perishable food supplies cared for?
❑	❑	Cups, glasses, etc., inverted on trays?
❑	❑	Sidestand top cleared except for allowed items?
❑	❑	Heating and other equipment turned off?
❑	❑	Soiled linen taken to soiled linen receptacle?
❑	❑	Traystands properly stored?
❑	❑	Proper amount of flatware, dishware, and glassware on sidestand for breakfast?

**Leave lights on for the inspection; turn off just before leaving.*

Credit card use is common. Servers must know what credit cards are honored by the operation. Some operations subscribe to some system of credit card verification that indicates whether the card should be honored. The server should use this system; if the card is verified, the server makes out the charge slip. The card with the slip is brought to the card owner who signs the slip, puts in the amount of tip, and receives a slip copy along with the credit card and a sincere thanks. On closing out the servers bank, these slips are offered instead of cash. Some operations ask the server to fill in a sheet indicating individual credit card charges and totals. A sheet reporting complimentaries, discounts, or other such transactions may also be required in the closing report.

At times, corrections must be made on ring-ups. An overring may be corrected by processing a void system set up by the operation. Similarly, a coupon or complimentary system sometimes requires that the check be adjusted to reflect the discount. Often, a ticket, card, coupon, or other item is offered by the guest to verify the right to the lower payment, and this is filed with the checks. If there is an overring, the server may have to write this on the check and also fill in the transaction on an overring sheet.

When no cash register is used, and the server carries the bank, discount and complimentary records are usually written up by the server on a record sheet. Servers must have an adequate mathematical and writing background to complete such reporting.

At the end of the shift, the register will give a total printout of transactions made by each server during the shift. The operation may also wish to have a form filled out recording the check number and check amount, with total sales given at the bottom. Servers must accompany such a form with the proper amount of money. When servers keep tips in the register, any overage in the register after the amount due on sales is set aside and belongs to the server. The server presents the original cash bank, the sale money, and other records to the appropriate authority for a signed release, and is freed of all further responsibility unless some discrepancy is found in the report or funds turned in.

CHAPTER SUMMARY

Mise en place—preparing ahead for the service period—is essential if excellent service is to occur. Time is saved and serving is organized. Prepreparation of needs should be planned and accomplished to cover all phases of a guest's visit, from the time of arrival to departure.

Grounds must be neat and clean. Good security and safety are important. Entrances, hallways, phone booths, coatrooms, and other areas guests encounter upon arriving require attention. Sanitation standards also are a must for todays service personnel.

Mise en place work needed to care for guests can be divided into three parts: (1) Getting ready. (2) Between-guests work. (3) Ending the meal. The first covers all preparations needed for the arrival of guests until they arrive at the dining area, and the work that has to be done before a meal starts, such as getting all items and condiments ready, getting checks, having a lineup meeting, and so on. The second part covers the activities that take place when one group of guests leave and the table must be readied for another. It requires fast work if the dining area is busy and guests are waiting for tables. The last part consists of the preparations that must be made for guest departure and the work that must be done to see that the dining area is ready for the next meal.

RELATED INTERNET SITES

Mise en Place while cooking
www.hertzmann.com/articles/2003/mise/

KEY TERMS

cash bank
mise en place
Grimod de la Reynière
silencer
traffic sheet

CHAPTER REVIEW

1. View several different kinds of facilities and look over the grounds, the entrance, and the way into the dining area. What do you find good and bad? What would you do differently?
2. What are some of the requirements of valet parkers?
3. How can a busperson assist the server in mise en place duties?
4. You are a server and have just arrived at the station assigned to you. What tasks should you perform, and in what order should you perform them?
5. What is a preshift meeting? What is its purpose?
6. How do some computers help stop security problems?
7. What is the proper way to clear a table of items before resetting it?
8. When guests are preparing to leave, what mise en place must be done to see that their departure is facilitated and they receive adequate attention to make them feel they were welcome?
9. Why shouldn't you wipe-dry wet dishes and glasses?
10. What system would you want management to have for the control of guest payment of checks?

Case Study

Good Mise en Place Is Missing

George is a new server in a section that during meal times is a busy place. This is his first service job since graduating from a school where service is taught as a part of the curriculum. The first day, he comes to work a bit early and is assigned a section of tables. He notes other servers around him are busy in their stations. He, however, studies the menu and goes to the kitchen to introduce himself to the cooks. He returns as the place opens up for the dinner hour. A few customers arrive and these tables are quickly waited upon by the other wait staff. George waits on his customers also, but he notices he's a bit busier than the others. The other servers seem to need to move around less than he does to serve the same number of guests. As the place gets busier and busier he feels a lot of pressure to keep up. He has to rush. He becomes frustrated and panics. Coworkers and the hostess have to step in and help him.

After the busy period is over and things are moving slowly, a woman coworker says, "You seem to know your stuff when it comes to waiting tables, but you sure don't know how to get ready to wait on them. Next time, you need to get ready. You didn't have your station prepared. If you had, you would have been able to serve a lot better and easier. We old timers know we've got to really get ready to make it through our busy hour. Next time I'd advise you to pack your station with things you'll need so that you can save steps when the rush period is on." George thanks her and says that the night has taught him a lesson. "I was told in class how important it is to do all your mise en place before and during the meal to provide good service. I guess I just forgot. Thanks for reminding me."

Plan a setup for George's station. Describe what he needs to do to get ready for the next time he serves.

SERVICE IN VARIOUS INDUSTRY SEGMENTS

OUTLINE

I. Introduction
II. Banquet Service
   A. Regular Banquet Service
   B. Banquet Table Service
   C. Receptions
   D. Showroom Service
   E. Kosher Meals
III. Service for Specific Meals
   A. Breakfast
   B. Lunch
   C. Dinner
   D. Tea
IV. Buffet Service
   A. Regular Buffet
   B. Cafeteria Service
V. Other Service
   A. Drive-through Service
   B. Room Service
   C. Counter Service
   D. Fountain Service
VI. Chapter Summary
VII. Chapter Review
VIII. Case Studies

LEARNING OBJECTIVE

After reading this chapter, you should be able to:

- Describe proper meal service and clearing for banquets, specific meals, buffets, and other types of service.

Introduction

The foodservice industry is one of the most varied in the world. This chapter offers an overview of how service is handled in many segments, from buffets to lunch counters to hotel room service.

Banquet Service

Regular Banquet Service

A banquet, whether a breakfast, lunch, or dinner, can be a special celebration, a gathering to honor someone, or a professional meeting, often preceded by a reception with beverages and hors d'oeuvres.

Many details go into managing a banquet. Some customers request dancing or entertainment between courses. This means there must be a way to provide music and a stage or floor for performing or dancing. Special flower arrangements may be requested. If there is a speaker or some type of program, the arrangements for this need to be agreed upon beforehand so everything can be ready. A sound system, photographers, special lighting, and a host of other items may also be needed.

The menu must be discussed and approved in advance, along with the total cost. The service style may be specified, or the facility may plan its own. At least one week before the banquet, the number of servers needed must be decided and an appropriate schedule should be created. Banquets usually require a special service staff. This must be taken into consideration when creating the service schedule. All servers should be aware of the proper dress or uniform for the event.

American and Russian service are the most common service styles used. In American service, food is placed onto the plates from which it will be eaten and brought to guests. In Russian service, all cooking, finishing, and carving are done in the kitchen. But dishing the food is done at tableside. Sometimes, a separate crew sets up the room with the bare tables and chairs. Some operations have their own employees lay the table linen and set the tables. Setup time can be worked into the service schedule.

A sample setting is usually provided so those setting up the room can follow the model. Often the setting up is a team effort, with some laying linen, others setting silverware and glassware, and others setting up salt, pepper, and other table appointments. Dishes and other items for the affair may be in a storeroom pantry. It is helpful and efficient if these items are on mobile carts so they can be rolled out to the dining area.

In standard functions, all flatware, china, and glassware are present. In more elaborate functions, the initial setup consists simply of a knife, a fork, and the water glass. In this case, before a course or wine is served, the proper silverware and glassware will be placed on the table. This may require a larger number of service staff and will increase the price for the event. When presetting all tableware there are two definite disadvantages: A large amount of tableware cramps the space, making it very uncomfortable for the guest, restricting freedom of movement and social interaction, and, when first seated, guests are often intimidated by seeing ten to fifteen pieces of flatware and countless glassware items in front of them.

In some banquet setups, coffee cups are preset. They should be positioned to the right of the guest but further up along the same line of the glassware so as not to cramp the space between the guests. The handle should always be at the three o'clock position. Generally, each cover needs at least 25 inches of linear space; if the cup is placed downward it will be too close to the bread and butter plate of the guest to the left.

Some managers do not feel that any food items, except ice water, should be on the table until guests are seated. Others may wish to have the first course, if cold, on the tables when guests enter. This may be done to speed service and may be done at the request of those in charge of the banquet. However, some operations even go farther. Not only will the first course be on the table but the bread and butter will be also.

A lineup or preshift meeting may also be used to give specific instructions on setting up. If the servers are accustomed to working banquets at the facility, the lineup

FIGURE 5.1

Banquet setup should be completed by the time guests enter the hall. Courtesy PhotoDisc/Getty Images

meeting may be simple with just the special needs discussed. If servers are new, then the meeting must cover procedures in detail with perhaps some repetition and emphasis to see that special needs are not overlooked.

Tables will be arranged as needed for the banquet according to a pre-arranged floor plan. Often there is a head table facing into the room. The speaker podium and other items will be set up at this table. Often this head table is on a raised platform, or dais, so it can be more easily seen. Various additional table arrangements may be used, but the most common is a 6- to 10-cover round table. Tables should be numbered, not only to let guests know where their table is, but also for assigning stations. Before guests enter the banquet room, it is important that the service areas are clean and presentable. A detailed check should be made to see that the arrangement of settings and tables is symmetrical and attractive. An individual in charge of service, and perhaps some assistants, are usually on the floor during the service to see that the function proceeds as planned and that every detail as specified in the banquet function sheet is handled.

Station assignments may be by individual server areas or by teams. When the **wave system** is adopted the entire banquet room becomes one large station. Once a starting point is established, the tables are served in sequence by all servers according to a preestablished order. Experienced banquet service personnel who have been working together for some time often prefer this method. Service can be synchronized so that while some servers clear the table after the various courses, others bring the next dish and serve it. It is customary to start service with the head table and to quickly follow this with service to the other tables so all guests start and finish eating at about the same time.

The wave system offers some advantages, but in a typical banquet situation, the individual station method is still preferred. A normal station for one server may include from 24 to 32 covers. This may be increased or decreased according to the complexity of the menu. The station that includes the head table may have fewer covers per server. In some areas the local union contract establishes the maximum number of guests to be served per server.

On average, the dining space required is 15 to 18 square feet per guest. The dining room layout depends on many different factors such as the type of event, the menu, the number of guests, the room shape, and the host's preference. Round tables are preferred in banquets as they are more suitable for conversation and more practical for service than any other shape. Sizes may vary, although the majority are either 60 inches in diameter (to accommodate six guests) or 72 inches (to accommodate eight to ten guests). Rectangular 6, 8, and 10-footers are also common. Square, oval, half round, and serpentine-shaped tables are used less frequently. Unless otherwise requested by the host, round tables are placed 4 to 5 feet apart to allow sufficient space for chairs and service traffic. Rectangular and other shapes would require additional space, normally 5 to 6 feet, although with the same space availability, a larger number of patrons can be seated by using rectangular tables as opposed to round ones.

For each table size and shape there is a corresponding table cloth; therefore, a 60-inch round table will require an 85-inch-by-85-inch cloth, while an 8-foot rectangular table (96-inch-by-36-inch) will be properly covered with a 54-inch-by-120-inch cloth. Special occasions and specific needs may require the various shaped tables to be placed in configurations such as a "T," an "L," a "U," an "E," or a square, just to name a few.

Since everyone eats the same meal at the same time, the service is simplified. This is helpful because it is essential that fast service occur. Usually banquets are planned on a tight schedule; often the meal must be over quickly so that the program to follow is not delayed.

Removing dishes after the program has started, or otherwise having server activity in the dining area when the program is going on, is very distracting and can lead to guest dissatisfaction with the whole function. Only if requested by the banquet host should the servers be allowed to walk around the tables or perform service duties during a speech, an award presentation, or any guest activities. In a typical situation, if a server is placing a tray of hot food on the jack, or stand, ready to serve, when a guest initiates an unscheduled toast or blessing, the server should wait for the guest to finish before continuing service.

Servers should learn the menu so they can plan their work, give smooth, efficient service, and provide competent answers to guests' inquiries about the items served. It would be a disappointing situation for the guest and an embarrassing one for the server if an answer could not be given in regards to the type of wine to be served, why it was selected to match various courses, what beverages are available, and so on. During the meal various other items may be needed. For coffee service at the end of the meal, cream and sugar may have to be brought to the table. Planning ahead and preparing these items is helpful. A complete mise en place, as discussed in Chapter 4, is essential for a satisfactory banquet service flow.

In some operations, the food items are dished up in advance and the plates are covered and placed over sheet pans on heated sheetracks on wheels, or similar holding containers. This method makes serving food a much simpler task for servers. Once the racks have been wheeled out to the dining room, the plates can be readily picked up and served. However, this practice may not be recommended if the food being served does not hold its appearance well. For example, after being in the various holding cabinets or sheet-racks for more than three minutes, sauces may develop a "skin" and garnishes can become soggy.

Dessert service may be spectacular or something simple. A special occasion cake may be cut ceremoniously, or simple sherbet or ice cream may be placed on the table. Depending on the specifics of the event, there can be a wide range of options for serving desserts.

If wine or any other special beverage is to be served, special pourers might perform this service so the regular servers can focus on serving the food. Whoever does the pouring should know the essentials of such service so it is done correctly. (Such service is discussed in Chapter 9.) However, some steps in traditional wine service are eliminated. The wine bottle is not shown to the host before pulling the cork, the cork is not pulled at the table, and no one is asked first to taste the wine to see if it is suitable for the occasion. These steps are unnecessary, as the host has usually specified and approved the wine selection prior to the event.

Those hosting the banquet must provide an estimate of the number of expected covers. A facility might allow a 10 percent overcount or undercount on such an estimate. Whenever guests must present a ticket or some other entrance permit, this may be used to confirm the count. Otherwise, a head count may be taken, but this can be difficult if the number of guests is large, such as greater than 2,000. Another way to confirm the number of covers is to have a plate count. To get a plate count, subtract the number of plates not used from the total plates set out. In this case, those hosting and paying for the banquet must accept the word of those making the plate count. Another way is to count the number of full tables and multiply by seats per table, then add the counts of the partial tables.

Banquet Table Service

As previously discussed, the two predominant service styles in banquets are American and Russian, with the American being the most popular. When serving courses according to the basic American service style rules, all foods are served from the left side with the left hand and cleared from the right with the right hand. Plates should be placed gently and attention must be paid so that they are positioned at least one inch from the edge of the table. All beverages are served from the right and cleared from the right. Left-handed servers might make some exceptions in clearing the plates if they need to handle dishware with a firm grip.

Once a tray of food is placed in the service stations, some servers like to pick up a maximum of four or five plates at once, as they feel they can service the whole station faster. It is recommended that no more than three plates be carried at one time unless they are small and "dry." When plates are ten inches in diameter or more and they contain thin sauces, the risk of spillage increases. Carrying a maximum of three plates is also more aesthetically acceptable. Serving one plate at a time is most desirable, but slows down service.

When pouring coffee, banquet servers often lift the saucer and cup from the table as they feel it is safer and less difficult. In reality the opposite is true. Coffee cups or any beverage container should not be removed from the table. To pour over it while the cup is resting on the table is actually safer as well as technically correct.

Breakage is a main concern of banquet operators, particularly during large functions. Banquet rooms have more breakage than any other type of foodservice operation. When collecting soiled dishes, items should be carefully separated and plates should be stacked according to size. This not only will minimize breakage but also speed up the breakdown procedures and will be of great help to the dishwashers.

Banquet servers normally carry trays on the left hand, flat, resting the tray weight on their shoulder. More experienced servers may carry heavy trays on the tips of the fingers. This should not be attempted by a novice, as it could result in breakage, embarrassment for the staff, and injury to a guest. (Refer to Chapter 7 for tray carrying guidelines.)

It is imperative that servers fully understand how crucial timing is in banquets. In the lineup meeting discussed earlier, the sequence of service should be addressed in great detail. Without proper planning, a smaller unforeseen incident might throw off the timing with disastrous consequences. The wise server always includes a minute or two for something unexpected in the sequence of service time flow. For example, in a banquet of three hundred guests where a prime rib dinner is served, guests usually will not question the cooking temperature of the meat. If a guest believes that the cut is too rare and asks that the meat be cooked a few more minutes, the request should be honored. One dish alone can disorganize an entire station, as the server has to make special accommodations for the guest. Meanwhile, the station is unattended and patrons are complaining. This is the time for other servers to step in and help. A fellow server can help keep an eye on the "abandoned" station. In some operations, servers play it by ear and proceed according to the individual station needs at any particular moment. Teamwork is essential to successful banquet service.

Buffet service should be used only with fairly small banquets. A large number makes for too much milling around and a lot of confusion. It is also difficult to move along quickly. If buffet service is used, the number of courses may be limited and often servers clear the table after the main course, then serve the dessert and beverage courses.

Receptions

Receptions are gatherings to celebrate special occasions or to honor a person or persons. They may be small, as for a small group of friends who are invited to meet a bride and groom, or they may be a large reception preceding a banquet. Often there will be a service bar or table at which beverages will be offered. Beverages may include punch, champagne, wine, beer, cocktails, and soft drinks. Nonalcohol beverages are becoming more common on reception menus.

If guests are expected to carry their beverages and hors d'oeuvres, servers will be needed to take empty plates, soiled serviettes, and beverage glasses from them. Tray stands with trays should be conveniently located away from sight, if possible, where these items may be placed before being returned to the kitchen for washing and sanitizing. However, instead of a bar or service table, servers might present trays of beverages and hors d'oeuvres. This is called **flying service** and the trays carried by servers are called **flying platters.** The Russians influenced the original French reception custom and introduced the concept of standing while eating.

Showroom Service

Showrooms often offer a very good meal and excellent entertainment at a nominal price. This attracts patrons to the facility, where gambling often occurs. The showroom with its food, beverage, and entertainment costs may lose money, but the loss is made up by the gambling department of the company. Service very similar to that of a showroom may also occur at a race track or other entertainment facility. Again, the price may be nominal, the purpose being to bring in patrons who will gamble on the races or otherwise spend money.

Sometimes, only beverages are offered, and this is most common for the late night show or for cabarets. Both food and beverage service and the entertainment are included in one cover charge. The cost usually includes a preset menu. Tips and additional items are extra. Showroom menus often offer a fairly elaborate meal with an appetizer, a salad, a main course that may be steak, prime rib, poultry, fish or seafood, along with a dessert and beverage. Beverage alcohol is usually not included, although in some cases a pre-dinner cocktail or after-dinner drink may be included in the cover charge.

Often a showroom facility has a large number of reservations and, when the door opens, these people are seated immediately. Seating 500 or more people within a few minutes is a challenge and requires good organization and planning. The service crew is usually experienced, working steadily at the operation for perhaps several years. If the operation is unionized, the union provides the required number of personnel. In one Las Vegas operation, as many as thirty five teams of two servers may be used to serve about 1,000 covers. (Today, servers in casinos tend to only serve cocktails during shows.) American service was the most common service used in these situations.

After the orders are taken they are brought immediately to the kitchen. Some operations have an announcer call out the orders to prevent confusion. This also allows servers time to do other things. Serving moves along rapidly since there are only a few pre-set items and these are often pre-portioned and ready to be placed onto the serving dishes.

Setup procedures for showrooms are the same as for banquets, with groups of servers working together to place all the necessary items on the tables. The servers are also expected to see that the required settings for the following show are ready at the end of the previous show. As soon as the show ends, servers must enter the room and quickly lay clean linen and set the tables. There is often a limited time before the next group of guests enters.

Kosher Meals

If your establishment serves *Kosher* foods, it will need to employ a Jew trained to oversee food preparation because it can be quite complex. It is common for people who don't usually observe the law of Kosher to serve Kosher foods at festivities and catered events, such as bar and bat mitzvahs and weddings.

All foodservice employees should be aware that the Jewish dietary law strictly forbids that meats, and all foods prepared with meat products, come in contact with dairy products. Kitchen or dining room equipment used for meat products cannot be used for dairy products. Kitchens used for kosher banquets have separate sinks, equipment, and different sets of working utensils, and often even two separate kitchens. This prevents the mixing of meats and dairy items in preparing, cooking, and serving.

Servers should become familiar with the basic principal's of *Kashruth,* the laws of keeping kosher. The term *kosher* means "fit" or "proper." All kosher foods must be handled, prepared, and served in full accordance with Jewish dietary law. This law allows only the consumption of meat from animals that have split hooves and chew their cud. Pigs have split hooves, but since they do not chew their cud, pork is forbidden. Only a specific list of birds, including most commonly eaten birds, such as chicken and turkey, are allowed. Only fish that have both fins and scales are permitted. This excludes all shellfish.

Service for Specific Meals

Breakfast

Breakfast service is usually faster than other meals because guests are often in a hurry, except at group gatherings where the occasion may be one where time is not so vital. Sometimes breakfasts are numbered on the menu so guests can pick out a group of foods and order them by number.

The normal American breakfast is often thought of as a fruit juice, cereal, entree item, bread, and hot beverage, but these items are not always present. Continental breakfasts are common, which include a fruit or juice, coffee or tea, bread or rolls, and butter. Buffet service is often used for breakfast service, with the buffet table holding fruits, juices, cereals, breads, and so on. Bacon, ham, and sausages may be also offered and a cook may be there to prepare eggs and omelettes to order.

There are a number of special-occasion breakfasts. A wedding breakfast may begin with beverage alcohol such as a gin fizz or champagne, or a fruit punch. A fresh fruit cocktail is often the first course, followed by a breakfast entree and bread. If the breakfast is late in the day, the entree may be quite substantial. A beverage such as coffee is served. Wine service may also occur.

A hunt breakfast is an elaborate buffet that may include a number of meats and other dishes. The meal is bountiful and heavy. A brunch is a fairly heavy breakfast served late in the morning. It is supposed to be a combination of the foods served at lunch and

FIGURE 5.2

Breakfast service can include both hot and cold items. Courtesy PhotoDisc Inc.

breakfast. Some brunches can be quite elaborate. A wide number of fruits and juices, meats, egg dishes, breads, salads, desserts, and other items may be offered. It is not unusual to see champagne served. The amount of service given at these occasion breakfasts may vary. If it is a buffet service, the server might only clear dishes and serve the beverage.

Lunch

The foods served at lunch can vary considerably. A salad or soup and beverage might be all that is desired. At other times a full lunch is wanted and this might be a first course, followed by a main entree and dessert.

The type of establishment and the foods being offered will determine the table setting. This may be a full setting even though the guest may order only a limited amount of food. Other operations will not have a complete setting but may bring the desired eating utensils after the order is taken. American service is most common for lunch service, unless the dining it quite formal and the luncheon is occurring at a club or fine dining establishment. In such a case, the menu might be quite elaborate with very formal service, but the tendency today is to simplify.

Group lunches are often a complete, but lighter meal consisting of a main dish, vegetables, salad, bread, dessert, and beverage. A first course, such as soup, may be served at a more formal meal. Often clubs will meet during the noon hour for lunch and hear a speaker. The service must be fast because the guests are often on a tight schedule. Normally, however, the entire luncheon may last one or two hours. Again, the cover setting should be appropriate for the items served. When groups meet this way, the service procedure is similar to that used at banquets.

Dinner

People generally have more time for dinner than for other meals and can use the meal as a period of relaxation after a day's work. Dinner usually includes a main entree, a starch item, a vegetable, and perhaps a salad. Bread and a beverage usually accompanies the meal. A dessert may or may not be wanted. People out to dine will often start the meal with an appetizer or soup.

People often dine out to celebrate an occasion and may want a bit more elaborate meal than they would normally consume. If this group eats in the regular dining room, the service is much like that given to other guests in the area. However, if the group meets in a special room, the service may take on some aspects of a banquet. It will all depend on the meal and the service desired. Servers who can do something to make the meal memorable are always appreciated.

Tea

The service requirements for a tea will depend on the type of tea and the number of people to be served. For small groups, the service can be somewhat like serving a family group at a meal, but for large groups, the service must take on the aspects of a banquet.

A tea may be just a quiet meeting of several or more friends who wish to get together and enjoy each other's company. In this case, service needs will be limited. The tea along with milk or cream, sugar or honey, and lemon slices will be brought, usually on a tray, and set on a small table. Cups, saucers, spoons, and small tea napkins should be provided on this table also. The server will pour the tea and bring it on a tray to each guest. The guest takes the tea and adds what is desired, taking also a napkin and spoon. If the group is quite small, the cups and other items along with the tray may be placed on a coffee table sitting between the guests and one will serve the tea and pass it. It is not unusual for a light cookie, tiny sandwich, or dessert plate to be included. It is important that servers see that guests are relieved of tea cups and other items as soon as they finish and also that the serving table is at all times kept filled with the necessary items.

There are low and high teas. The plainest low tea is simply tea with sugar or honey, milk or cream, and lemon. A more elaborate low tea might include cookies, mints, bonbons, or nuts, or may offer a dessert or fancy sandwiches. Low teas can be seated service or standing. Large group teas are almost always standing. A high tea is like the most elaborate low tea, but some heavier foods may be offered, so in a sense it becomes a light meal. (The British often have high tea late in the afternoon and then have a late dinner.) High teas are always seated service and are usually not practical for large groups.

Buffet Service

Regular Buffet Service

A buffet is a service concept that presents foods on a table. Guests progress along the length of the buffet serving themselves, or indicate what they desire to servers attending the tables. They vary considerably in the kinds of foods offered, when they are used, and in the table service required to serve guests.

A common buffet arrangement is one in which guests serve themselves vegetables and side items and then progress to the end of the buffet where a carver carves and serves the meat item or hot items and places them on the guest's plate. For a breakfast buffet, instead of carved meat the server may make waffles, hot cakes, or egg dishes. Following this method, the hot food is served last, immediately before the guest sits down to eat. Some buffets are arranged so only one kind of food is on one table. The main course may be on one table, the vegetables on another, desserts on another, and so on. Guests go from table to table as they please. This prevents long lines. Such a service is called **shopping service.**

A banquet may offer a sit-down dinner, then an elaborate buffet of different desserts, which is called a **sweet table.** In some buffet service, guests are provided a tray and a tray rail on which to slide their dishes, instead of carrying them in their hands. This latter method is the one commonly used in cafeterias where guests get everything they need and bring it to the table. Tables may or may not be set with flatware. In some cases, again like cafeteria service, the silverware may be rolled in the napkin for pickup by the guests. The serving side of the table is often protected by a small, clear glass panel over the foods. The glass is high enough for guests to reach under and get the food, but it protects food from any germs spread by breathing or sneezing. Because of this, the panel is called a **sneeze guard.** Hot dishes are usually heated by electric lamps. Cold dishes may often be on a refrigerated tabletop. If only hot food is served on a plate, then the plate temperature should be warm, but not too hot to hold safely. Likewise, plates used for cold food should be chilled. If both hot and cold foods are served on the same plate, then the plate should be at room temperature.

Buffets have both advantages and disadvantages. Certain buffets can handle a large group of people quickly, but the menu must be limited and the service time short. Thus, a hotel housing a large tour group that would be difficult to serve using traditional table service may be able to use buffet service for the group. Buffets also appeal to guests who are in a hurry, allowing them to select what they want and eat immediately. Buffets also require a smaller service staff than other service concepts. Unfortunately, buffets lead to a higher food waste because guests often take more than they can eat. In some cases, servings may be controlled by proportioning or other portion control methods.

It is extremely important that the foods on a buffet be attractively prepared and presented. Food of different heights can be arranged to give a variation or different items can be placed on a raised platform. The tasteful use of floral arrangements, mirrors, and fabric can add to the presentation. Colors should be bright, natural, clear, and varied. Garnishing can be somewhat elaborate, but not garish or overdone. Some operations use a buffet to feature an elaborate brunch on Sundays or holidays.

Three types of ethnic buffets are also used to attract trade or give a special emphasis to an event: a smorgasbord, a Russian buffet, and a French buffet.

The smorgasbord originated in the Scandinavian countries. It offers a full meal with many kinds of hot and cold foods, desserts, and beverages. To be a typical smorgasbord, pickled herring, rye bread, and mysost or ejetost cheese should be offered. A typical hot offering is Swedish meatballs. A Russian buffet is also an elaborate offering of hot and cold foods. There is often a fairly large offering of roasts and other cooked meats, including game. The buffet should offer rye bread with sweet butter and caviar in a glass or ice carving bowl. Vodka will also be served.

A French buffet is a fairly elaborate offering that begins with a course of hors d'oeuvres and appetizers. Patrons select what they want, and servers take it to their seats. The main buffet usually offers a fairly wide array of roasts and substantial dinner offerings. Again, patrons choose their food while the servers take the items selected to their seats. Wine is usually served as well as coffee or other beverages. A dessert table may be featured, or desserts may be arranged on voiture (carts) and brought to seated patrons for selection. French buffets are not common.

Recently the **action station** concept of buffet service has become increasingly popular. It consists of cooking or finishing some food items on medium-sized pans over portable rechauds as the guests go through the buffet line. The foods items are served by the server from the pan or to the guests' plate Russian style. Most patrons are very receptive to this type of service as they feel their food is freshly cooked, as opposed to the common belief that buffet food may be allowed to sit for long periods of time. The action station concept can vary from a basic buffet line where a vegetable and chicken stir-fry or a pasta is offered, to a more elaborate setup where items such as shrimp scampi or beef tenderloin medallions are served to the guest standing in line. If the latter is adopted, servers must have a thorough knowledge of menu and preparation techniques. The equipment, tools, and ingredients needed are the same as in any tableside preparation methods typical of upscale restaurants. Obviously, this increases considerably the service level and the overall cost of the meal charge.

Cafeteria Service

Cafeteria service is a type of buffet service where the guest selects the items desired and then carries them away to be consumed elsewhere. Foods are arranged in the same sequence as a menu in a typical restaurant, therefore appetizers are usually first, then salads, pastas, vegetables, main items, and desserts. Servers behind counters serve the foods after the patron makes a selection. Many foods may be preportioned, such as appetizers, salads, and desserts, and these will be selected and picked up by patrons. Patrons pick up their own eating utensils and beverages using the self-service facilities.

Often the operation offers more than one counter in order to avoid long lines. A shopping center arrangement where different groups of foods are obtained from separate sections has been explained previously. In some cafeteria systems, different ethnic foods or regional specialties are obtained from different counter locations.

Payment for the foods selected is often made at the end of the selection counter at a cashier stand. Servers or buspersons will clear soiled items after guests leave, clean the table, and prepare it for the next patron. Occasionally they will also provide service on special requests.

FIGURE 5.3
Cafeteria service is one type of buffet service. Courtesy the U.S. Department of Agriculture

Other Service

Drive-through Service

There are two kinds of service where food and beverages are ordered from a car: drive-through and car-hop. The latter is seldom seen today because of the increased labor cost over drive-through service, but it was the most common type used in the early days of take-out food.

Drive-through service is relatively simple. Customers drive to an ordering station where a menu is displayed. They use a two-way communication system to place their orders. They continue forward to the service window, pay for, and receive their order. The car drives away, leaving the space for the next car. Customers can park and eat in their car, using the handy trash containers. Or they can take the food home to eat with their families. Often people on the road will stop and get food and eat as they are driving. This is particularly convenient for families with small children. An additional advantage of drive-through is that during rush hour, when there is a long line inside the restaurant, service at the window is still reasonably fast. Many quick-service operations have extended their business hours, and a great number of patrons prefer the drive-through, particularly at very late hours, for reasons of safety.

Traditionally, management gives somewhat of a priority to drive-through customers. Recently, some establishments have expanded the traditional system of two stops into three—one for ordering the food, one for payment, and one for pick-up. Operators say the additional window allows for quicker service and helps maintain sanitation standards so that employees do not have to count money and handle food (or items that come in contact with food) at the same time.

Drive-through employees should be convivial and say good-bye while giving the food order to the customer. Using phrases such as, "Have a nice day," "Please drive carefully," and "Come back and see us again," will leave the customer with a pleasant feeling.

In this fast, mobile, and vibrant society of ours, drive-through food operations are increasingly popular. Some quick-service operations report that as much as 79 percent of total sales come from the drive-through section of the business.

Room Service

Hotels and motels often offer room service, in which guests select foods and order them by phone. The food selections are then delivered to the guests' room.

Breakfasts can be preselected the night before by marking a menu left in the room and hanging it outside the door on the door knob. It is picked up during the night and the breakfast is delivered the next day at the desired time indicated on the card. Delivery carts for room service should be preset and ready to go, so there is little delay. Some carts may have a unit on them that keeps hot foods hot and cold foods cold. Usually, however, covers are placed over the food to keep it at the desired temperature. Before leaving the kitchen, the server should check carefully to see that everything needed is on the cart, including all sauces, condiments, and utensils. If something is missing when the food is delivered at the room, the guest will not be able to eat immediately. Often the guest places the order by phone. The average time span between the ordering and the food delivery should be 15 to 30 minutes. Certain menu items might require more time to prepare. Often a special service elevator is used to deliver such foods.

FIGURE 5.4
Room service is a popular amenity at many hotels and motels. Courtesy PhotoDisc/Getty Images

Servers should check before knocking on the door. The room number is written on the guest check. After knocking, the server should identify himself or herself by saying, "Room service."

Once the guest opens the door, the proper approach is to address the guest and offer a sincere smile. When entering the room the server should always be preceded by the cart. In some operations, the server has the option of setting the table in the room and placing the food items on it, leaving the tray with the food on it on the table, or leaving the cart in the room allowing the guest or guests to serve the food. Some room service carts are equipped with folding leaves. Once they are opened and locked into position, the cart becomes a full-sized table. Because of the nature of this type of service, some guests are sensitive to privacy and expect the server to leave as soon as possible.

The server will ask the guest if there is any other request and will present the check. More than 90 percent of guests prefer the check to be included in the final room charges as opposed to paying cash. If the check does not indicate that gratuities have already been included, it is customary to tip the server directly, according to standard tip rates.

Counter Service

Good counter service is not easy. It is fast work, and servers are often under considerable pressure. The turnover is short, multiplying the number of settings, orders, servings, and clearing. Guests are usually in a hurry and want fast service, and counter service is perfect for this purpose.

A counter station normally ranges from eight to twenty covers, and the work area for the server must be carefully planned so the service can go smoothly. Everything needed for settings and service should be within reach. At many counters, a place mat is used and the cover setting is placed on this with the knife and spoon to the guest's right and the forks and napkin to the left. Water is placed at the right above the tip of the knife. If other items are needed, they are placed at the cover after the order is taken. Condiments should also be close so all the server has to do is reach for them. Tote boxes or bus pans for soiled dishes may be close to the server's work area so the server does not have to go far to deposit soiled items. It is customary for buspersons to bring clean items and remove soiled ones. They may also help clean counters and maintain the area in other ways. Service can be speeded if the place where orders are placed and food is picked up are close to the counter area. If not, some way of getting orders into the kitchen without the server having to go there can be coordinated.

After a guest has been given the check—with a thank you—and has left, the soiled items should be quickly removed and a new setting put into place. A new guest may have already taken the seat expecting fast service. The clearing and cleaning of the counter top and placing of the new cover setting should be done with smooth, efficient motions.

In spite of the pressure of work, servers should remember to be cheerful, friendly, and considerate of guests' needs. At times when a counter server is not busy at the counter, side work should be done. If napkin holders are used, these should be filled, condiment tops should be wiped clean, salts and peppers set, and if need be, filled, along with sugar bowls. A failure to do good mise en place during slow times can make for a much more difficult and frustrating experience during busy ones.

Standup counter service is sometimes used to give fast service. Most often patrons select their food or beverage at one station and then carry the items to the counter, and bring what is needed. Often only fast, short-order foods are served in this style. A fountain or even a bar may be a part of such service.

FIGURE 5.5

Ice cream parlors are still a treat for children and adults alike. Courtesy PhotoDisc Inc.

Fountain Service

At one time ice cream parlors were very popular. They went into a decline several years ago, but now are beginning to regain popularity. Today the product usually comes in 2 1/2–gallon paper containers and is prepared under good sanitary conditions. A much wider range of ice cream flavors and other frozen desserts are also available. The most common fountain items are sundaes, sodas, floats, milkshakes, splits, and carbonated drinks. The facility will usually have standard recipes outlining the proper way to make them and the correct ingredients and their amounts. Soft ice cream is very popular, as well as frozen desserts made from yogurt or nondairy products. In addition, there are sherbets, ices, frappes, and milkshake mixes. The frozen product also can be kept in units that produce and hold items at freezing temperatures. Spoons are the most common eating utensils used for fountain services, and these are placed to the right of the cover. It is usual to see paper napkins used.

CHAPTER SUMMARY

There are different styles of service, such as American, French, or English (butler). These service styles are explained more fully in Chapter 7. These various styles are used in different service concepts such as buffets, counter, cafeteria, or banquet. Although both are called a service, they are somewhat different in meaning. Sometimes service concepts are divided into occasion style of service—such as banquet, reception, tea, or breakfast—or into place style of service, where the service is distinctive to a place or location like a counter, showroom, or cafeteria.

Occasion services discussed in this chapter are banquet and reception. Place services discussed are buffet, cafeteria, counter, fountain, room, and showroom. Banquet service occurs usually at a dinner where a group gathers together to eat a common meal in celebration of something. The service style is usually American or Russian. Table service includes the kind of service given at breakfast, lunch, or dinner. Teas may be low or high—a low tea being simple with just tea and perhaps one or two simple food items, and a high tea being much more formal with a larger offering of foods that make it almost a meal. Receptions are almost always standup affairs. Service may take place from a buffet or servers may carry trays of beverages or small finger foods and offer them to guests. The trays on which such items are offered are called *flying platters*, having received this name from the Russians who invented this style of service.

Buffet service is perhaps the most common place service. It is a style of self-service with foods and beverages offered at a buffet, counter, or table. Some buffets, such as Russian, smorgasbord, or French, offer a distinctive type of food and beverage. Cafeteria service is another self-service distinctive to a place. Items are picked up at a counter—some are selected by customers and then served by servers behind the counter—and take to a table (or out of the operation) and consumed.

Counter service is a fast type of service where the guest is seated at a counter and orders the items desired, which are then brought and served at the counter. Good server organization is required because the service must be fast. Fountain service is a special type of counter service. Frozen desserts are featured, but beverages and other items may also be served. Room service is used in places where guests stay in their club, motel, or hotel rooms and order food. Servers must see that everything needed is on the delivery cart since the items are served far away from the place where they are prepared. Delivery time must be short so that hot foods remain hot and cold foods remain cold.

RELATED INTERNET SITES

Beverage Services
www.beveragenet.net

KEY TERMS

action station
flying platter
flying service
shopping service
sneeze guard
sweet table
wave system

CHAPTER REVIEW

1. Define style of service, the occasion service concept, and the place service concept.
2. What is a banquet, and what styles of service are commonly used with it?
3. What are a smorgasbord, a Russian buffet, and a French buffet?
4. What is the difference between a high tea and a low tea?
5. What is a reception?
6. What is a flying platter?
7. What is a buffet?
8. What are some of the variations that can occur in cafeteria service?
9. What is room service, and what are some special challenges it poses?
10. How is the food order placed in drive-through service?

CASE STUDIES

Managing a Menu

You are planning to open a light-meal restaurant a block away from the University of Wisconsin's Madison campus. Patronage is expected to be largely students. Students attending classes in the day are expected to be a good source of revenue, but after the late afternoon classes are over, meals for students living in the dorms and nearby university housing are hoped to be an additional source of revenue. Evening meals generally cost more than midday food, so the revenue split is estimated to be about 50–50 even though patronage is less for evening hours. The operation is designed to be a place with some refinement in the food offerings and service in order to compete with the university food services.

Plan a menu for this operation. Design it to meet both the budget needs and food preferences of students. Give thought to student dietary patterns. Some students will be vegetarians. Others will want low-carb or low-calorie foods. Most students have to watch costs. They are also becoming better informed on what's included in an adequate diet and want to follow it. Most like sweets. Portions can be large; they are young and have healthy appetites. It might be that on evening menus some items should be offered in large, medium, and small portions with appropriate pricing for portion size. Decide if this would also be a good idea for the lunch menu.

The Roast Beef Incident

Lynn Moore was waiting on a frequent customer who had brought guests. She took the orders. One lady ordered roast beef, medium rare. Lynn went to the kitchen and placed the orders. The cook said the evening was late and the roast had stood in the steam table so long it had cooked to medium. Lynn was tired

and did not want to go back to the dining room to discuss order options with the customer. She decided to take it medium, thinking the lady would accept it anyway. Her experience was that many people do.

When the orders were ready, she brought them and served them. The lady got her beef medium. The lady said nothing, and Lynn thought it would be O.K. but no. The host said, "Lady, please finish serving so we can eat while our food is hot, but take Mrs. Wright's order back and bring it back medium rare." Here, Lynn made a second mistake by replying, "Sir, I brought the lady what she ordered." "That's not medium rare, it's medium," the host insisted, "Take it back and get the correct order." "It *is* the right order, sir," the server said. The argument continued getting hotter and hotter while she served the rest of the orders. By the time she finished serving the orders, she knew she was in real trouble.

At this point, Mrs. Wright interposed, protesting to the host, "Paul, Paul, it's all right. I've tasted the beef and I find it quite acceptable, nice and tender, juicy and flavorful. Please let it go." But the host was too angry, and continued dressing Lynn up and down. Lynn was near tears and tried to explain her mistake and apologize, but the host was by now too angry and cut her off with, "I don't want to hear your excuses. What I want you to do is to bring Mrs. Wright medium-rare roast beef!"

Lynn now said abruptly, "Excuse me," and left the table for the manager's office. Lynn told the manager the story, admitting her mistakes. Together they returned to the table, where the manager said to the host, "Sir, I'm the manager. I understand there's something wrong with an order. Can I be of help?" The host explained, almost choking in anger. "Sir," the manager said, "my server made a mistake. She should have told you our roast this late in the evening was only medium. What I'd like to do for our error is to compliment you with the meals and also give every guest all the wine they want with their dinner. I also apologize for our error."

The host suddenly relaxed, saying, "Oh my! That's nice of you! You didn't have to do that. But thank you. Maybe I made a mistake too, getting angry and refusing to listen to the server. I'm sorry for getting angry." The manager then turned to Mrs. Wright and said, "I'm sorry, we're out of medium-rare roast beef and all we have is medium. Can I bring you something else?" "Oh, no," Mrs. Wright replied, "I find it quite delicious." She had eaten over half of her order.

The problem solved and the table quiet, the manager and Lynn left and a new server came.

Admittedly, Lynn's two falsifications were mistakes that caused the problem. How should she have taken care of the problem when she originally placed the

order? If not then, when she brought the orders, what should she have told Mrs. Wright about her order? What should the host have done when Mrs. Wright said the first time the beef was acceptable? Was the host at fault in any way? Was the manager right in making such a generous offer? Could he have tried giving less before offering what he did? Was the manager right in giving the table a new server? Why? What did he accomplish by changing servers? What should the manager do to Lynn? Discharge her, or do something less stringent? (Lynn was a good employee; guests liked her and some guests when they arrived even asked for her tables. She consistently was among the leaders in sales and in generating tips.)

SERVICE AREAS AND EQUIPMENT

OUTLINE

LEARNING OBJECTIVES

After reading this chapter, you should be able to:

- List and describe the equipment typically found in the dining area of an operation, as well as items used in table service.
- Describe the traditional hierarchy of a service staff.

Introduction

Servers use many kinds of service equipment and should know what these items are called and how they are used. Each cover, or a guest's place at the table, must have the proper equipment set in the proper way. Different service styles require different equipment.

Menus are a means of communication between guests and servers, and servers should know the items that are on them. Servers also work with two distinct staffs: the kitchen staff and the dining staff. They need to know the organization of each so they can work with them.

Dining Area Equipment

Personal Items

Servers should always carry an order pad or order checks, with a pen or pencil. Very formal service requires a serviette or service napkin. Matches or a lighter can be carried to light candles.

Service Stations

The service station, or wait station, assists the service staff in performing their duties. Depending on the type of the operation, a service station may be small with only a minimal amount of dishes and flatware, or it can take up a large portion of the floor space and have many types of equipment.

What is stored in a service station depends on your type of operation. Some items found in service stations include flatware, dinnerware, table condiments, various paper goods, hot and cold prepared foods, glasses, cups, linens, trays, and tray stands.

All service stations perform one or more basic functions:

1. They store basic goods so the server does not have to make numerous trips to the main storage area for each customer's request.
2. They hold items for cleaning the dining room.
3. They act as command and communication centers where orders can be placed with the kitchen and management and the kitchen can communicate information to servers.
4. They serve as a production center.
5. In some instances, they serve as host/hostess stations.
6. They act as a link between the servers and kitchen staff.

In a banquet operation, the service station might be as simple as a table along a wall with extra silverware, napkins, bread and butter, and water and coffee pitchers. In a large chain quick-service operation, it might take up 50 percent of the floor space, including the area between the counter where the customers order, and the pass-through area where the kitchen places prepared food before it is brought to the customer.

Storage Center

The service station, as a storage area, varies greatly from one operation to another. At the most basic level, the setup is simple and the costs are minimal. However, larger operations have service areas that are more complex. They sometimes contain temperature storage units that hold both hot and cold prepared foods. The most common cold storage machines are ice bins, soda dispensers, and milk machines. They are found in most operations. Cold storage might also include dessert cases, or small reach-in refrigerators that hold premade salads, garnishes, or side dishes. Hot or warm storage are usually soup wells and bread warmers. In both hot and cold storage, the item is completed earlier in the kitchen, and the servers may have to plate or bowl the item before serving the customers.

As the costs of building restaurants increase, operators try to decrease the amount of space that does not directly have contact with the customer. Because of this, floor space in service stations is decreasing. To accommodate this change, restaurants are using vertical space—the space from the ceiling to the floor—to expand the usable space in service stations. It is important that managers are sure that proper sanitation and safety procedures are followed.

Cleaning Area

The service station is sometimes used for storing cleaning items for the dining area. In some operations, the service station includes a place for dirty dishes/glasses, used linen, and garbage cans. Most also have sanitizing solutions to clean a table after it has been bussed. These solutions are used to clean a table after each use so it may be used, or *turned*, again.

Service stations or an attached area usually include the items needed for major cleaning after meal periods, at the end of the shift, or at closing. These include vacuums, polish, and cleaners.

It is management's job to establish strict, high sanitation rules for its operation and also see that employees have read and understand them. Management should see to it that all employees go through training programs on sanitation procedures. Many localities require that employees take an approved course, and are certified after successful completion as sanitarians qualified to supervise the kitchen and/or dining areas.

The established rules should apply to all areas of the operation, but special rules may apply to specific areas to fit special needs. The rules applying to kitchen and dining areas should be the same, except the rules may have to also cover sanitation for guests in non-kitchen areas. As an example, most states require that soups and other liquid or semiliquid foods be heated to a sufficiently high temperature in the kitchen as to guarantee the destruction of all microbiological agents (as opposed to being heated in a steam table). This is only one of many state and local sanitation regulations that must be followed.

It is important for safety reasons to keep used eating utensils, dishes, glasses, and used linens away from food and clean items. Some states require a hand washing sink to be available in the kitchen and service areas and that this sink be used only for that purpose.

Command Center

In many foodservice operations, the service stations act as a command and communication center. POS (point of sale) systems allow servers to place orders to the kitchen in an orderly and traceable way. The POS system not only records and informs the kitchen of the customers order, but prompts needed information such as appropriate temperature and side dishes, tracks open checks, counts customers, totals check sales, and monitors amount of sales. The service station is also where management and the kitchen communicate important information to the service staff. There might be a board that lists the soups of the day, specials, or what the kitchen might be out of. Management can use it to inform the service staff of schedules, section assignments, or special circumstances. In most operations, the service station allows the staff to communicate with each other.

Production Center

It is incorrect to assume that all production in a foodservice operation takes place in the kitchen. Many operations now require their service staff to be involved in some aspect of production. It might be as simple as making coffee and tea, or as complicated as preparing an entree-type salad. Beverages are most commonly filled in a service station. Almost all operations require the service staff to make coffee. Iced tea and lemonade are other simple beverages often produced by the service staff. Operations that offer malts and shakes usually require the servers to prepare them. With the growth of premium coffee and specialty drinks, the skills level and training required of the staff has increased. In operations that are food oriented but offer beverage alcohol, the servers might perform bartending duties. The necessary equipment must be located in the service station.

The production of food in a service area is usually limited to cold foods that are prepared beforehand, but do not hold well in their final state. These include ice cream, desserts with whipped cream, and salads. All the needed ingredients are prepared earlier by the kitchen staff or the server and are assembled as ordered to ensure the highest-quality product for the customer.

Host or Hostess Station

FIGURE 6.1

The host/hostess station is often the first interior area where guests have first contact with the staff.

All operations have an area where the customers first have contact with an employee. Most often this is a host/hostess station area. Most stations have a few basic items; phone, reservation book, menus, and a floor diagram (including open tables, server areas, and smoking/nonsmoking sections). Some have POS systems that allow them to function like a cashier. Depending on the operation, it might combine some functions of a service station.

Kitchen or Service Area 

The kitchen or service area is in direct contact with the kitchen. Usually a manager or an **expediter** is in charge of the area. The main function this person is to act as a communication link between the servers and kitchen staff (with the assistance of or instead of a POS system). If servers need to communicate with the kitchen, they tell the expediter, who relays the information to the kitchen. The expediter is also used if the kitchen needs to talk to a server about clarifying an order. Most operations that use an expediter require the service and kitchen staff to communicate through the expediter. They also help servers with tray orders and assist in any production the server might have to perform.

Furniture

Tables 

Tables for seated service are used in most operations. In others, counters with stools are used. Stand-up areas may only have high counters at which people stand while eating. Tables are usually sized to accommodate different-sized groups, but may be put together for larger groups. Some square tables have hinged leaves that can be lifted up and locked into place to make a larger table, square or round. Table arrangements are dictated by serving needs and the dining area shape. Whatever the arrangement, it should be symmetrical and neat. Many tables are equipped with levelers; servers should know how to use these.

It is important that tables be far enough apart as to permit easy passage of people going by. Also, guests like tables to be far enough apart as to not allow a low conversation to be heard at an adjoining table. Aisle space between tables will be less wide than aisle space for guests walking through to get to other areas of the room.

Chairs 

Chair sizes and shapes will vary, but all should be very sturdy. Usually, the chairs do not have arms. Chairs should be large enough so guests can sit comfortably. Chairs on rollers are easy to move, but may be dangerous because they can easily slip out from under a guest as they are about to sit or stand up. Many chairs are equipped with glides so they move easily over a surface. Servers should check these to see that they are secure. Also, they should check to see that the chairs are clean. Some high chairs may be needed for small children, while booster seats for larger children might be needed. Chairs should be checked to note rough edges or legs that can snag on guests' clothing.

Other Furniture

A **guéridon** is a small cart often carrying a **réchaud,** a small heater on which one cooks or reheats foods and liquids. The cooking items needed are on the cooking top or on a shelf under. The dishes on which to serve the food are usually kept on a shelf under the food-preparing top. Carts called **voitures** are used to move hors d'oeuvres, appetizers, salads, fruits, desserts, beverages, wines, or liqueurs around the dining area. Offerings should be neatly arranged. Some voitures may be refrigerated, while others may be heated and hold roasts and other items for carving and serving. A bus cart is used by servers or buspersons to hold used wares until they are transported to the warewashing area. Most operations do not allow the cart to be brought into the dining area when guests are present. Usually it is placed in an inconspicuous place and used wares are brought to it.

Tables are usually covered first with a silencer—a felt pad that quiets the noises of dishes and utensils and absorbs spills—and then with a tablecloth. Cotton, linen, and ramie are the most commonly used fabrics for tablecloths. Tablecloths should fit the table and hang down only about eight inches. They should be plain and smoothly pressed. The proper placement of silencers and tablecloths is described in Chapter 4.

Fine-dining operations may use naperones to cover the tablecloth. These are square, and do not cover the entire table.

Napkins, about 18 to 20 inches square, are often used in fine-dining establishments for dinner. Smaller ones may be used for lunch or breakfast. In more casual dining, paper napkins may be used. Small cocktail or hors d'oeuvre paper napkins may be used at receptions, cocktail parties, or teas.

A serviette or hand napkin is usually made of cloth and hangs over the arm of the server to protect his hands when serving hot items and as a general serving towel. Serviettes should be used in all kinds of service situations. Placemats are individual covers, about 18 inches to 24 inches long and about 12 inches wide. They are made of various materials, and can be decorated with advertising or information the operation wants a guest to read, or games for children. Some placemats have the menu printed on them.

Table Service Equipment

Dishes

Dishware costs money and servers should give it good care. A dinner plate can cost $12 or more. Some operations prominently display the cost of individual dish items in the dishwashing area. A server who drops a tray of dishes costs the operation—or themselves—a lot of money. Sliding dishes and glassware from a tray onto a dishwasher table should never be allowed. Keeping flatware (forks, knives, spoons, etc.) separate from dishes is a must; some operations have a special sink or container where flatware can be dropped to soak before washing. Glassware should be given special care and kept separate from dishes.

Soiled ware may be loaded into dish baskets or tote boxes and taken to or from the kitchen. Servers who come to an area carrying a very heavy basket with no place to set it down can cause breakage or accidents. Know there is a landing space before you carry. Areas where soiled items are brought should have sufficient space so crowding and cross-traffic are avoided.

Most operations use a heavy reinforced china (porcelain) for dishware. Some may use lighter china, but this is often too fragile and expensive for ordinary use. It cannot stand the heavy wear and handling. Others may use ceramics or plastic, and some use paper. Whatever is used should be sanitary and clean. All china should be sparkling clean and sanitized when placed before guests. Glassware or china that is chipped, cracked, or has lost some of its glaze should not be used.

A description of some of the most used dishes follows.

Platter — Platters are long, oval dishes used to hold several portions of food; some smaller ones may be used instead of plates. Sizzling platters are almost always made of metal and are heated to a very high heat so the food sizzles on them as they are taken to the guests. Often an underliner is used. Guests should be warned not to touch them when the platter is set down.

Service/show plate — In fine-dining operations, a plate larger than the dinner plate sits on the table at the cover when the guest is seated. Hors d'oeuvres, soups, and other first-course foods in their serving dishes are placed on this plate. In many cases, management instructs servers to immediately remove the plate after the cocktail is served. Once removed, show plates are not used again. Show plates are usually 11 to 12 inches in diameter. They can be expensive and should be handled

	carefully. After their use, they should be wiped gently with a service napkin wet with vinegar and never brought to the dishwasher.
Dinner plate	This is used for dinners and even lunch to hold the main part of the meal. It can also be used as a service plate for other purposes, and is usually 10 to 11 inches in diameter. In many establishments, oval shapes are used instead of the traditional round plate.
Salad plate	A 7- to 8-inch-diameter plate used to hold salads, soups in bowls, desserts, and other foods.
Bread plate	Often called a bread and butter plate, it is usually 4 to 5 inches in diameter. Used for breads, rolls, toast, butter, jellies, jams, and so forth. Can be used as an underliner with a doily over it.
Demitasse	A round container with handle holding four to six ounces of hot beverage; a saucer is placed below the cup. Primarily used to serve espresso and specialty coffees.
Finger bowl	A cuplike container holding lukewarm water, usually with a slice of lemon or a rose petal floating in it. Used for lightly rinsing or washing the fingers. It is placed on an underliner with a doily on it. It may also be called a *monkey dish.*
Cocotte	Small casserole for cooking and serving entree preparations. It often has a cover over it.
Cloche	A bell-shaped cover used with a serving dish; the cover holds in the heat. Can be made of glass, plastic, or metal.
Snail dish	Metal or ceramic stoneware with six deep indentations for holding snails cooked out of their shells. A shallower one with six indentations is used for snails in the shell.
Sauce boat	For serving gravies, au jus, and other liquids.

Glassware

All glassware should be sparkling clean and sanitized. Often an all-purpose wine glass is used in place of various wine glasses. However, interest in the beverage selections can be increased by using a variety of specialized glassware.

Service Utensils and Flatware

All foodservice operations have special utensils that are only used by service staff. They can be as simple as plastic tongs for portioning salads, or as formal as sterling silver serving sets. The most common utensils are ice scoops, tongs, ladles, stirrers, and spatulas. The utensils your servers need will depend on your operation. In any operation, the utensils must be in good shape, clean, and sanitized.

A list of commonly used utensils follows.

FIGURE 6.2
Behind the bar, glassware should also be clean, organized, and easily accessible.

Dinner fork	Used for main entrees and other foods eaten from a dinner plate. Can also be used as a general utility fork.
Salad fork	For salads, appetizers, some desserts, or fruits.
Fish fork	Used for fish and sometimes seafood dishes.
Cocktail fork	Used for seafood and other cocktails.
Lobster fork	For lobster when served in the shell.
Dessert fork	For pies, cakes, pastries, and other solid desserts.
Oyster fork	For eating clams, oysters, and other bivalves.
Fondue fork	A long fork used to pick up bread cubes and dip them into a cheese fondue; a shorter fork holds meat in hot oil for a meat fondue.
Teaspoon	Used for eating vegetables, fruit sauces, puddings, fruits, etc., and other foods that are difficult to eat with a fork.
Tablespoon	Larger than a teaspoon, used for soups or cereals.
Soup spoon	Used mainly for soups.
Coffee spoon	For beverages, some cocktails, and ice cream.
Espresso spoon	For liquids served in a demitasse cup.
Sundae or iced tea spoon	For ice cream sundaes, ice sodas, tall iced beverages, or similar beverages served in deep glassware.
Sauce spoon	A wide, shallow spoon used for sauces and lifting foods out of casseroles.
Snail tongs	For holding snails in the shell so the snail fork can extract them.
Lobster tongs	Used for holding lobsters in the shell.
Pastry tongs	For picking up and serving pastries.
Cake or pie server	For serving cakes, pies, pastries, tortes, and similar desserts.

Other Items

Some other items commonly used by servers are pepper mills, water pitchers, coffee servers, teapots, bottle openers, oil and vinegar cruets, mustard holders, bread baskets, cheese graters, service trays, tray stands, check holders, and guest caddies.

MENUS

Menus are important to service because they inform the guest what your operation offers. Most operations have menus in readable form. This may be a traditional individual menu, a sign listing specials, or a board above the counter. Some operations require servers to recite what the menu offers. This limits the menu size, but adds a certain atmosphere.

It is usual to print menus on hard paper called *bristol* or cover stock, but they can be on a sign board as used in a cafeteria or a drive-in. In informal restaurants, a placemat may hold the menu. Some operations have the menu on their order tickets.

Meal plans and menus differ. A meal plan lists the type of food served at each course of the meal, while the menu lists the exact items served. **Exhibit 6.1** shows this difference.

Menus may be designed for specific meals or occasions. Thus, we can have a breakfast, lunch, afternoon, dinner, or evening menu, a room service menu, or a banquet menu. Some may even be designed for the person, such as a children's menu, an early bird menu, or a vegetarian menu for guests with special dietary needs. Some menus are not intended to be used for ordering by guests, but only to inform cooks of preparation needs.

Menus are sometimes named by their nature. An **a la carte menu** prices all menu items individually. A **table d' hôte menu** is one that prices foods together in a group, often as a nearly complete meal. A **du jour menu** (*du jour* means of the day) lists foods available only that day, and often there is little choice in selection—many times, only one du jour meal is offered. A **cycle menu** is one that runs for a period of time with foods changing daily. Then the cycle is repeated. A **California menu** is one that lists snacks, breakfasts, lunches, and dinners all on one menu. A **general menu** is the main menu of a hospital from which special diets are planned. Thus, from the general menu the dietitian selects food that a specific patient might have.

It is customary to list foods on a menu in the order in which they are usually eaten. Others may even have a separate menu for appetizers or desserts. Although less usual, menus may list different food groups by courses.

EXHIBIT 6.1 A Meal Plan and its Menu

MEAL PLAN	MENU
Appetizer	Melon au Kirsch
Soup	Lobster Bisque
Entree	Baked Ham, Champagne Sauce
Side dishes	Sweet Potato, Roll, Harvard Beets
Salad	Pear and Grape with Honey French Dressing
Dessert	Hazelnut Cream

Menu items often carry some designator indicating how they are prepared. Thus, the menu may list a grilled cheese sandwich, which means it is browned on both sides on a griddle. It is not toasted. However, the word *grilled* used with a steak usually means the steak is broiled over or under direct heat. A menu listing Prime Rib of Beef is not saying the meat is of prime grade, but that the cut comes from the area known as the *prime* area of beef. *Roasting* means baking in dry heat, while braising means cooking in a small amount of liquid. *Boiled* means cooking in a lot of liquid. *Florentine* means the item comes prepared in some way with spinach. Servers should know what terms of this type mean. If they do not know a menu term they should look it up or ask the chef or manager, so when guests ask they are prepared to give an accurate answer.

EXHIBIT 6.2 Some Common Preparation Times (cooking times only)

BROILED STEAKS, LARGE SIRLOINS		CHICKEN	
Rare	20–24 minutes	Broiled or sautéed	20–45 minutes
Medium	28 minutes		
Well done	28–32 minutes	FISH AND SEAFOOD	
		Filets	6–10 minutes
LAMB CHOPS		Steaks	6–16 minutes
Medium	12–25 minutes	Whole	13–20 minutes
Well done	14–30 minutes	Baked oysters	12–16 minutes
		Steamed clams	20 minutes
PORK CHOPS		Boiled lobster	20 minutes
Well done	15–30 minutes	Broiled lobster, split	15–20 minutes
		Broiled tails, split	8–10 minutes
BARBECUED RIBS			
Well done	45 minutes		

Servers likewise have to know the time it will take to prepare orders so they can inform guests on approximate waiting times and also plan their service time schedule for that particular table. **Exhibit 6.2** gives some preparation times, but these are averages, and may vary from the preparation time of a particular item on a menu based on the facility in which one is serving.

The times in Exhibit 6.2 can vary according to cooking method. As servers gain experience, they begin to almost intuitively know about the time needed for menu items. However, if this experience is lacking, it is best to check.

Beverages, when offered on menus, are listed in various ways. Banquet or fine-dining menus often list the wine to the right of the menu items with which it is served. Or the beverages served with the meal may be listed on the bottom of the menu:

Oysters Rockefeller
Boston Bean Potage, Croutons
Crab Mousse, Bercy Sauce
Grilled Loin Lamb Chops, Mint Sauce
Orange Glazed Carrots
Baked Stuffed Potato
Oriental Salad with Shiitake Mushrooms
Baked Stuffed Apple

Cocktails
Moet and Chandon Champagne
Cousino Masul Reservas Cabernet Sauvignon—1988
Chateau d' Y' chem Sauternes—1988
Liqueurs

Some operations may have a separate menu listing all beverage alcohol, usually aperitifs, cocktails, spirit drinks, brandies and liqueurs, after-dinner drinks, beers, and nonalcohol drinks. The wine list usually separates domestic wines from foreign ones and in the order of listing in each is usually aperitif wines, dry white wines, dry red wines, sweet dessert wines, fortified wines, sparkling wines, and alcohol-free ones. Wines are often numbered and coded on the menu, and guests can then order by the number. The server also lists on the check the wine by its number or code.

Service Staff

Two staffs are used to work in food preparation and serving: the production or kitchen staff, and the service staff. Servers are in frequent contact with the production staff where the type of staffing impacts what servers do in ordering and picking up orders. A thorough knowledge of kitchen staffing can help servers move more smoothly in the kitchen.

Kitchen Staff

Many food services operate with only one cook on a shift, along with a helper or two, and a dishwasher, and even the dishwasher can be missing. The cook usually directs the other kitchen workers, but frequently receives direction from an owner, manager, or assistant manager. If there is more than one cook, one is usually designated as head cook or chef, and this person is in charge, working under the direction of a manager. Kitchen staffs can grow to a considerable size with departments such as baking, salad, vegetable preparation, and cooking. Each department usually has a head that directs the work in that department; in many, the overseer of the kitchen may be a kitchen manager or chef. In some health facilities this may be a dietitian; dietary aids may assist the dietitian and also work closely with cooks in food preparation and in the dishing of foods and beverages to see that the foods meet the nutritional needs of the guests.

In the typical French organization there is often a continental kitchen. An executive chef is in charge, and is the overall manager of all food preparation and functions associated with it. A steward works with the chef in ordering, storage, and perhaps in menu planning. The steward is also in charge of much of the dining equipment, nappery, and other dining room equipment.

In food production, the *sous chef* (second to the main chef) runs the kitchen for the executive chef, who usually has an office some place away from the kitchen. The continental kitchen is divided into various departments headed by *chefs de parti,* such as the *garde manger* (cold foods and pantry), *patisserie* (bake shop), *legumier* (cooked vegetable and garnishes), *potager* (soups and stews), and *entremetier* (sauced and roasted dishes). Under these chefs de parti will be assistants and helpers. The great chef Auguste Escoffier introduced this continental type of kitchen organization. He wanted to keep his kitchen quiet, and so instead of having a number of servers calling orders to various units, servers brought their orders to an *aboyeur* (announcer) who, after receiving an order from a server, called it out in a clear, loud voice so the cooks who had to prepare the items could hear.

Some kitchens use checkers to make sure orders are correct, to price orders, and to give them to the aboyeur to call out. It is the duty of the checker to check the food and beverages leaving the kitchen against the order to see that everything is correct, even the garnishes. In this manner, servers take from the kitchen only the correct order, nothing else. Some checkers act as cashiers, receiving money for the orders and giving back the proper change.

In addition other helpers, sanitarians, and employees will be on the staff. Servers do not come into contact with many of these employees. There are frequent opportunities for conflict when working under stress, and friendly, cordial relationships among employees can do much to smooth over times when the work is highly demanding. Servers and kitchen staff work under the pressure of getting orders out promptly and in the standard required. A large number of orders may come in so fast that they pile up, and the kitchen staff may be under high pressure to get the food out in time and in proper quality. This is no time to make special demands. In some cases, servers may share tips with cooks; this often creates a high degree of cooperation between cooks and servers.

Traditional Service Staff

Service staffs vary considerably. All servers are usually responsible to someone who heads the operation or a representative of management. Management hires, sets up schedules and stations, sets up service standards and training, and directs other service work. In larger organizations, some of these duties may be delegated.

Some operations such as drive-ins or quick-service operations have little or no staff organization. Orders are taken at a counter or window, and the food is brought there to give to guests.

Seated service operations are more complex. Usually a manager or some representative of management is in charge, with one or several servers working on a shift. If the service staff is large, a hostess or host may be in charge.

The most complex staff is modeled after that used in Europe, which grew out of the staffs used by Escoffier and Ritz. It is labor intensive and expensive, but elegant and lavish. The manager of serving is usually called *maitre d' hotel,* but may be called host or head waiter. Often this person is in charge of assigning stations and may even hire the serving staff. Under this leadership may be *captains* who are in charge of a group of servers, often called *chefs de rang,* who may also be called front servers. Each chef de rang is assigned a station numbering up to 25 seats. A *commis,* also called assistant waiter or back server, assists the chef de rang. Buspersons in this organization are often called *commis debarasseur.*

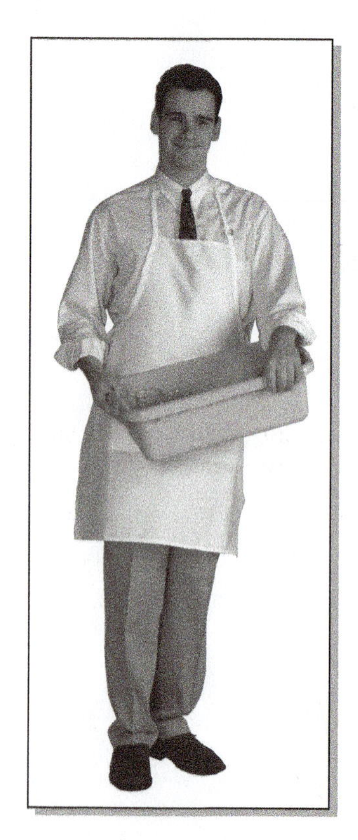

FIGURE 6.3

A service busperson assists servers and clears the table once guests have left. Courtesy Corbis Digital Stock

In some French server staffing the chef de service acts as the maitre d', and may have an assistant called the *chef d' étage* who directs serving in the dining areas. An individual called the *maitre d' hotel de care* supervises a dining room section, somewhat as captains do.

Other persons who may be included in the serving staff are the food and beverage manager, who is in charge of all food production and service. Banquet managers head up banquet catering staffs and are usually under the food and beverage manager. The wine steward, or *sommelier,* is responsible for wines and other beverage alcohols and their service. Some units use bartenders who only fill orders for servers, but usually do not serve the public. Service staff are discussed in more detail in Chapter 7.

In almost any type of service buspersons are used to bring trays of orders to a station, take away soiled items, and assist servers.

In a healthcare facility, a dietitian is usually in charge of service, but nurses who are not accountable to this person often deliver the food after it is sent from the kitchen. Servers may serve in visitor's or doctor's dining rooms.

CHAPTER SUMMARY

Servers work with a lot of equipment, and to serve properly they must know what this equipment is and how it is to be used. In some operations the amount of knowledge required is much less than in others. A server working in a restaurant serving French food will use a wide variety of equipment, while one working in a drive-in will use a very different variety of equipment.

Each server needs certain personal items such as an order pad, pencil or pen, and other equipment needed to take orders and complete the serving of them.

Servers serve guests seated at tables or booths. These should be clean and neatly set. Service stations hold much equipment needed by servers. Servers should see they are properly stocked and are kept clean and orderly. In some food services mobile equipment may be used, and this also should be neat and clean, and not left where it can interfere with the work of serving. Other equipment will be used according to the serving needs and type of operation.

Tablecloths, napkins, naperones, placemats, and other napery may be required, and servers need to know how to use these so that tables appear neat and well maintained.

Some servers have to handle a wide assortment of dishes, glassware, and eating utensils. The use of glassware or dishes with certain eating utensils is prescribed for certain foods or beverages, and servers must know the proper ones to use. Different dishes, glassware, or eating utensils require specific placement at covers. Without a thorough knowledge of all the factors that go with the proper use of these items, good service cannot occur.

Menus are used to communicate to guests the items available and how they are prepared, along with the price. Servers need to know what the items are and how they are prepared so they can completely inform guests about menu offerings. Servers may have to explain menu items and also indicate preparation times. Servers have to be sales people as well as do the work of serving.

The highest of sanitation standards must be kept by servers who inter a with guests. The sever, area, server stations, tabletop, items on the table, and menu should be immaculate when guests arrive. Dishes and other tableware should be handled so that no placewhere food might rest is touched by the server.

Servers need to know to whom they report and the role other employees on staff play in accomplishing food production and service. They need to know the kitchen organizations so they know how to place orders and pick them up. They also need to know this organization so they can work with the food preparation staffs.

Related Internet Sites

Virtual Trade Show

This site links burgers and sellers of food and beverage industry products.

www.nightclub.com/virtual_new

Key Terms

a la carte menu
California menu
cycle menu
du jour menu
expediter
general menu
guéridon
maitre d' hotel
réchaud
table d' hôte menu
voiture

CHAPTER REVIEW

1. What is a service station? What is its use? What does it normally hold?
2. Match the terms on the left with their definition on the right.

_____ (1) Guéridon	a. Small casserole
_____ (2) Naperone	b. Used with a dish having small indentations in it
_____ (3) Snail tongs	c. Covers tablecloth
_____ (4) Demitasse	d. Holds about 4 ounces of espresso
_____ (5) Cocotte	e. Used to hold food and equipment for tableside preparation

3. Match the terms on the left with their definition on the right.

_____ (1) Aboyeur	a. Assistant waiter
_____ (2) Maitre d' hotel	b. Front waiter
_____ (3) Commis debarasseur	c. In charge of a group of waiters
_____ (4) Commis	d. Busperson
_____ (5) Chef d' étage	e. In charge of wine service
_____ (6) Sommelier	f. Assistant to chef de service
_____ (7) Chef de rang	g. Calls out food orders

4. What is a cycle menu?
5. In what order are items usually presented on menus?
6. What kind of menus are not meant to be used by guests for ordering?
7. What is cover stock?
8. Explain the difference between a *guéridon* and a *voiture.*
9. What is the difference between an *a la carte* and a *table d' hôte* menu?
10. What advantages and disadvantages do you see in servers sharing tips with the cooks?

Case Study

Server Stations

You are assisting an architect who is planning a kitchen, and the architect asks you to design the service stations. Design the service station she should use as an individual server station. Also design one that can act to supply the needs of three or four servers.

Classic Service Styles

Outline

I. Introduction
II. French Service
   A. Tossing and Mixing
   B. Plate Presentation and Sauces
   C. Deboning and Carving
   D. Flambéing
   E. Fire Plan
   F. Serving Hot Items
III. Russian Service
IV. American Service
V. English Service
VI. Chinese Service
VII. Chapter Summary
VIII. Chapter Review
IX. Case Study

Learning Objective

After reading this module, you should be able to:

- Describe the techniques of French, Russian, American, English, and Chinese service styles.

Introduction

Serving food and beverages to guests formally can be accomplished in different ways, often referred to as *service styles.* All authorities agree on the basic names and definitions of the many styles, but the details of each service style can vary with each authority. This module strives to give the most useful and agreed-upon information.

After all, the service style used by a facility should be the best suited to the facility's needs. Using an inflexible style is unwise when some change better suits the needs of the operation. The original formal style was developed to best suit the needs of the originating operation. Thus, while the rules of various styles will be cited, the authors in no way will give them strong support in view of others who do not agree.

Although service style is the way food is presented to guests, a service concept is the type of service offered to guests. The five classic service styles used are French, Russian, English, American, and Chinese. Alternate names are sometimes used: *cart service* or *guéridon service* (French service), *plate service* (American service), *platter service* (Russian service), *family-style service* (Chinese service), and *butler service* (English service). Although there are as many service styles as there are cultures and nations on earth, this chapter confines itself to the five most traditional, or "classic" styles.

French Service

French service evolved from the elegant and lavish style of the French court, **le grand couvert** (the great cover). It has been considerably modified, however, and today many operations have introduced so many variations that it is difficult to indicate just what it really is. The distinctive feature that marks French service is that the final preparation of the food is at tableside and served from the service table to guests. Thus, the food comes from the kitchen in an unprepared or semiprepared condition, possibly in a sauté pan, on a cart. Since final food preparation occurs where guests can see what is done, service is very much a show. Considerable skill and finesse are required to do it properly.

French service is the most labor-intensive of all formal styles, and, therefore, the most expensive. It is also the slowest, reducing the number of guest turnovers in the facility. Because of the large number of items used in tableside cooking, warewashing costs are increased. More space is required so fewer tables can fit in an area. More experienced and skilled servers are required. The menu price because of all these factors must be high. It is perhaps the most elegant, and only a small number of diners are willing to pay the extra cost.

As discussed in Chapter 6, the table on which preparation occurs is called a *guéridon.* The heater used to cook the food is called a *réchaud,* which means to rewarm.

FIGURE 7.1
In French service, the final preparation of food takes place tableside and is then served to the guests.

The guéridon contains the necessary tools and equipment to perform the required work. Not all food served from the guéridon has to be cooked. A salad might be mixed on it, or fish may be deboned, or a piece of meat might be carved on it.

The main, or front, server is called the **chef de rang**, who is assisted by a **commis de rang** and perhaps a busperson, called a **commis debarasseur** (*debarass* means to clear or take away). In addition to preparing the food, the chef de rang dishes and garnishes it. Warmed plates are used for hot food. The prepared plate is then handed to the commis, who serves it to the guest. The commis debarasseur assists the commis and the chef de rang. They traditionally wear black ties and white gloves. The chef de rang might wear a white apron when at the guéridon.

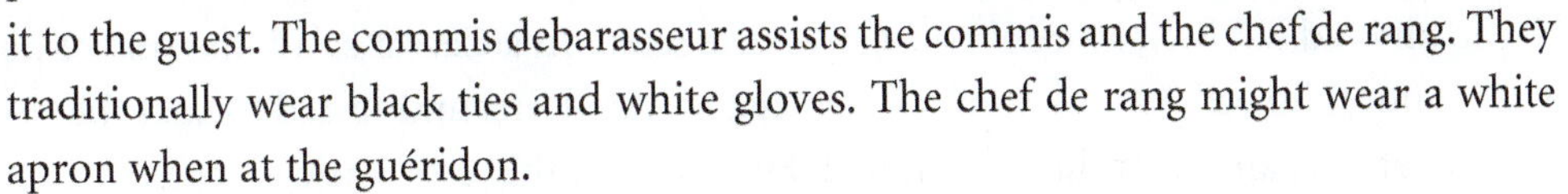

A dining room captain is the direct supervisor of the team and sees that the service proceeds as desired. At times, the more difficult preparations may be done by the captain, and often the captain may debone or carve. Occasionally, the maitre d' might lend a hand in more intricate preparations. It is not uncommon to see this individual flambéing, since this can be a dangerous procedure.

Alcohol and other beverages are usually served by a cocktail server, but wine service is provided by the **sommelier** (wine steward). In some operations, the *chef de rang* takes orders for beverages before dinner and brings drinks to the table and serves them. However, the sommelier, if one is on staff, still serves the wine.

The typical setting for French service varies. No more than three pieces of flatware appear on either the right or left side of the cover. Only three wine glasses are permitted. If more flatware or glassware is needed, they are added as needed. Salt and pepper are rarely on the table. In some operations, no bread and butter plate is used. A service or show plate is sometimes in the center of the cover, which is used as an underliner until the main course is served. A napkin is usually on the service plate. No ashtrays will be on the table; it is not considered proper to smoke until the meal is completed. It may then be allowed, according to the rules of the local government or the establishment.

Soup is brought in a tureen by the commis de rang to the guéridon, then is placed on the réchaud and dished into warm soup plates. Sauces are often finished by the chef de rang with much of the basic preparation being done in the kitchen.

Coffee in demitasse cups may be served after the meal. Orders for after-dinner beverages are often taken by the chef de rang and brought to the table and served. Even if no finger bowls are used at other parts of the meal, the meal usually ends with their service.

There are some special techniques that are commonly done by the chef de rang. These are briefly described below.

Tossing and Mixing

Tossing and mixing is used primarily to make salads. Because guests are near the guéridon, it is important to toss or mix by working toward the server. Thus, if there is any spattering, it will be toward the server and not the guests. It is also important to handle utensils and tools quietly and efficiently. Do not scrape or hit the bottom or sides of the bowl. Mix silently and gently, using quick, short motions.

In putting on the dressing, it is best to add oil first, so the salad is coated with it. Next vinegar, lemon juice, or other liquid ingredients can be added. Seasonings and croutons are added last. If the oil is added after any liquid, the salad pieces are then coated with moisture and the oil tends to run off and down to the bottom of the bowl. If the oil is added first, more is retained on the salad. In a Caesar salad, the rule of oil first is not followed. Instead, the coddled egg is added first and tossed to coat the romaine lettuce, and then the oil is added. However, in this case, the egg tends to capture and retain the oil, so the desired effect is achieved.

Plate Presentation and Sauces

It is important when dishing foods to place them in the center of the serving dish so they do not flow over the plate rim. The plate rim should act as a frame surrounding the food. Always garnish attractively.

Sauces are common in French cooking. In some cases, the bottom of the serving dish will be covered with the sauce and the item placed on top of it. Or the sauce may be poured over the item, covering or partially covering it. Another way is to serve the sauce on the side, in a sauceboat or gooseneck. Lemon is often served with fish and shellfish. Often this is a lemon half wrapped in fine cheesecloth so the seeds do not squeeze out with the juice.

Deboning and Carving

It takes experience to debone foods well. The server needs to know the bone and muscle structure of each item. Fish have bony fins and a skeletal structure in which the belly bones are attached to the spinal bone. Poultry has a *keel* (breast) bone, and legs and wings attached to a backbone. The bone and muscle structure of meats depends on the cut. Many meat cuts will have much of their bone structure removed before cooking. Thus, a rack of lamb or a prime rib will have the chine bone (spine bone) removed, leaving only the ribs. A leg of lamb or other cut may have all bones removed, so carving is merely a matter of slicing the meat.

It is important in deboning fish to leave the deboned parts whole. The remaining bones left on a plate may look messy. First, remove the fin bones. There will be one usually in the middle of the back, several on the belly, and one near the tail. Place a fork into the fish and using a spoon's edge, pull the fin bones away from the fish. If necessary, the fork can first be moved around the fin to help loosen it and then the spoon can easily pull the fin bones from the fish. Next, with the fork, run down the back under the flesh and lift up. The top fillet should be easily lifted up in a whole piece. Lay it on the serving plate. Carefully remove the meat covering the belly bones and lower tail. Next, by lifting with the fork at the tail, remove the spinal column from the other half of the fish. With it should come the belly bones, leaving the under fillet ready to place on the plate. It is important to see that all bones are completely removed. Some are very tiny and hard to see. Some fish, such as trout and salmon, have a ridge of bones that do not come with the spinal column. In large fish, these are easy to see and remove. If the trout or other fish is quite small, these bones are so small that they can be eaten with the flesh.

In deboning, it is critical to use the correct knife and proper utensils. To carve well, one must have good tools. Two basic items are a proper knife for the job and a carving fork to hold the item securely while being carved. A 10-inch carving knife is needed to slice roasts and cut through a rack of lamb, and a 7-inch narrow, rigid knife is needed for poultry. It is important to use the knife and fork to handle the meat during cutting and plating. This ensures the best safety and sanitation standards. Always use a sanitized carving board.

Most meats are sliced across the grain. The carver should note the grain and turn the meat so the slice is across it.

Some poultry, such as a duckling, is split in two by laying the cooked bird breast side down on the carving board, and, using a sharp butcher knife, cutting through the back and breast. The keel bone is removed and one half is served as a portion. It is not usual to debone a chicken entirely. First, the wings are removed by cutting through the wing joint at the breast. It is common to remove the wing tip and leave only the two wing joints. These may be separated in large birds, but in smaller birds, they are usually not. They are not deboned. Next the legs are removed by pulling the leg away from the body and separating the leg where it joins the back at the hip joint. The thigh and leg may be separated in small birds and are usually not deboned. Now, by cutting down closely along the ridge of the keel bones, the two breasts can be removed whole. Any slicing of the legs and breasts can then occur. Sometimes with a large bird such as a turkey or goose, no deboning is done. The meat is sliced from the bird as it sits on a platter, breast-side up. However, the wings are removed before such a carving process starts. The legs may not be, but some carvers prefer to remove them, and then, holding the bone of the leg, slice off the meat.

In Europe, some of the traditional fine dining establishments use a **chef de trancheur,** or carver. This person rolls in a cart (*voiture*) with various roasts, such as game, lamb, tenderloin, or poultry, which are carved according to guest selection of the item. The person must be highly skilled to perform this fine, old tradition.

Flambéing

Flambéing is most often done in French service, but today flambéing is not usually done at tableside because of the danger of accidents. However, if the person doing the flambéing is experienced and competent, the presentation can be spectacular and very entertaining for guests. It is most important that no bottle of alcohol be allowed to come near a flame. If this happens, the vapors of the alcohol at the mouth of the bottle can ignite, causing the bottle to explode and burst into flames.

A person can catch fire and be killed or horribly burned and disfigured. Only, repeat only, an experienced, skilled and practiced employee should be allowed to do it, and everyone should know what to do in case of an accident. Flambéing does not involve cooking. The item with the flammable material is warmed over a heater—usually a candle, alcohol, or gas flame. Keep the flame low; never let it be so high as to reach the sides of the container.

The equipment required is a réchaud, and blazer or sauté pan. Some shallow copper pans are used for items like crêpes. The amount of alcohol added to the sauce must be balanced. Too much sauce overwhelms the alcohol and it may fail to ignite or burn poorly. The right amount will just cover the surface with a thin film. Warm it slightly so it reaches a temperature at which the alcohol begins to rapidly vaporize. It is these vapors that ignite. To ignite, tilt the dish slightly toward the flame. The vapors should ignite and the flame rapidly spread over the surface. Have a towel or a serviette nearby that can be quickly thrown over the flame should any accident occur. This towel will smother the flame by not allowing it to have oxygen. Do not serve the item while it is still flaming. Wait for the flames to die out and then serve. Desserts like Crêpes Suzette, Bananas Foster, Baked Alaska, and Cherries Jubilee are all common flambéed desserts.

Fire Plan

A fire plan should be established and every employee should know its content. It should instruct employees how to protect patrons and evacuate them. A plan might look like this:

1. See that a fire protection unit, either one from within the building or the fire department, is called. If everyone must evacuate, immediately call from another location as soon as possible, Don't assume someone else will do it unless you see someone actually doing it.
2. Immediately take what steps you can to put out a controllable fire. Grab any material nearby, such as flame-retardant draperies, to throw over a fire.
3. Know where fire extinguishing materials are. They should be kept nearby and the easily accessible.
4. Turn off all possible sources of flame, such as a gas line.

5. If evacuated, check to see that everyone is accounted for, such as all members at one's tables, and are safe and away from harm.
6. Follow any instructions of the plan.

Serving Hot Items

Servers delivering hot food to patrons should always warn them if the food or plate might burn someone. Often, steaks are delivered on a sizzling hot metal platter. Soups can be the cause of burns. Servers need to use care when serving things such as hot coffee or tea so they do not drop the contents on those they are serving. They should not allow a patron to hold the container when filling it with a hot liquid. They should hold it themselves when filling it and set it down, or fill the container setting in its proper place so it does not have to be handled. Do not fill containers to the rim.

Russian Service

Russian service originally came from the French, but was so greatly modified that it became a distinct service style. All food preparation is done in the kitchen, and the food is ready to serve when it leaves the kitchen. It is less labor-intensive because only one server is used, and is faster than French service. It also requires less space, but is well suited to the seated service of large groups eating the same meal, and can be quite elegant. Guests, as in French service, feel they are getting personal service because of the way they are served.

In Russian service, all cooking, finishing, and carving are done in the kitchen. Only the dishing is done at the tableside. Liquids such as soups and sauces are placed in tureens and are brought to the table. Sometimes soup is brought to the table in small, individual bowls. In Russian service, food items are placed from a platter onto a plate directly in front of the guest, customarily from the left side.

FIGURE 7.2
In Russian service, only the dishing and presentation is done at the table.

Usually hot food is piping hot; so hot that servers can burn their arms where the platter rests on it. Because of this, many servers wrap a small towel around their arm that helps insulate them against heat. (A jacket sleeve hides the towel.) When serving is done from a platter, the server faces the guest and rests the platter on the left arm (for a right-handed person) and left-hand palm. If it is a smaller dish, the dish rests only on the palm. This requires considerable skill to balance the item and then serve from it. Portions of food are lifted from the service dish by the right hand by placing a large spoon with the bowl facing up under a portion, and a fork with tines up is placed on top to hold it in place. A spoon is used to serve vegetables and many other items.

The table setting for Russian service is basically like that for French service, but an ashtray, water glasses, and bread and butter plates might also be included. Since Russian service is a bit less formal than French, the rules are not as strict.

Dishes on which food is to be served are set down in front of guests by the server, using the right hand to place the item on the right side of guests. The server then moves to the next guest on the left and sets down the proper item, and then moves around the table clockwise until all have their dishes.

Coffee is served at the end of the meal, with cups and saucers brought to the table and the coffee then poured. Cream and sugar will be brought, if desired. Finger bowls are not usually served in Russian service, but this can vary.

American Service

American service is the fastest, least labor-intensive service style. It is the most commonly used, and can vary from somewhat formal to casual dining. The food is dished onto the plates from which it will be eaten and taken to the guests. This saves time and assures that hot food is hot and cold food is cold. Because most of the food of the meal is served and garnished on one plate, fewer dishes are used, reducing usage and warewashing costs. Because the food can be plated and garnished in the kitchen under the supervision of professional chefs, the food as it comes to the guest can be very attractive. The space required for such service is also minimized. Many chefs welcome American service because it gives them the opportunity to display their creativity by arranging individual food items in a manner that is visually appealing.

FIGURE 7.3

The server presents the plated food from the guest's left in American service.

American service can also be modified easily to suit many situations and needs. Thus, it can be varied to suit breakfast, lunch, or dinner service. French service is difficult to modify in this manner. Russian service is less difficult, but still causes some problems.

The standard place setting is knife with blade facing inward and then spoons on the right and forks on the left. The tips of handles of flatware should be about a half inch from the table edge. Normally, it is proper to set spoons according to the progression of courses. This places the soup spoon often on the far left, but it is not considered wrong to place the teaspoon on the outside and the soup spoon next to it. The amount of flatware used varies according to the menu, but the limit should be three on either side. If more are needed, they are brought at the time the items are served. A service plate may or may not be used. In its place may be an hors d'oeuvre plate. A bread and butter plate may also appear. It is placed above the forks and sometimes slightly to the left. The butter knife should be set on the bread and butter plate at a right angle or parallel to the forks.

FIGURE 7.4

The American table setting is the most widely used table setting in food service. Courtesy PhotoDisc, Inc.

The water glass or goblet is to the far right where glasses are placed above the spoons and dinner knife. Wine glasses are placed left of the water. Ashtrays, salt and pepper shakers, and decorations are typically in the center of the table, or against the wall in a booth. The coffee cup and saucer may be on the table just to the right and slightly above the knife and spoons. Coffee is sometimes served with the meal in American service. Cream and sugar are usually on the table at the start of the meal, but may be served later with coffee. In some fine dining establishments, the coffee cup and saucer will not be on the table but will be brought to the table just before coffee is served. Chairs should be out from the table away from the tablecloth, if one is used.

The table may or may not be covered with a tablecloth. Napkins may be folded in various ways. Usually they are placed in the center of the cover in front of the guest.

Except in very informal situations, service moves around the table from right to left. Normally, bread is served in a basket, except for formal occasions when it might be served by the server, who uses tongs to place the bread on plates and a fork to serve the butter or margarine. It is also common to see butter on a small plate on the table. On a large table, several bread baskets may be set down along with several plates of butter or margarine.

The general rule in serving is to serve the main entree and other items that go in front of the guest from the left, using the left hand. Other items such as water, coffee, and wine should be served from the right using the right hand. Salads, bread, butter or margarine, and other items are set from the left using the left hand. These rules and the others given here are, however, frequently violated, and, if it improves the service, there is no reason why rules cannot be changed. Most rules of service should be established on the basis of what is best for the guest, and easiest and quickest to do for the server.

It is not proper to remove any dishes or start a new course until everyone has finished at the table. In busy banquets or other occasions this rule is violated. The server takes advantage of opportunities to remove soiled dishes and avoid having to take more time later to remove them. Soiled items are removed from the table by removing items from the left-hand side with the left hand and items on the right-hand side with the right hand. Crumbing may be done in more formal meals. **Crumbing** is removing all particles of food left on the table after the entree, in order to have a clean table before the dessert is served.

An average of about fifteen covers is a good work load for a server for a dinner using American service; never more than twenty-five. Often, team service works well with American service because it makes possible the use of several servers to proceed with dispatch in one area. While the back server is picking up the food, the front server can be available for guests' special requests.

English Service

English service, also called butler service, uses less labor than Russian, and is a little bit faster. It is rarely used in the United States but is seen in some clubs and private parties. It has a graciousness desirable with more intimate groups.

Often the first course is on the table when the guests enter the dining area. Iced water will also be poured and butter or margarine and bread may also be on the table, although bread may not be brought until needed. Soup may be served dished into individual bowls or cups by the server, or the soup, in a tureen, may be placed in front of a party's host or hostess who then serves. This person might hand the soup to the nearest guests, who pass it along to other guests, or to the server who will then take it to the guests. Salads might be served this way or brought already dished, and served by the server.

FIGURE 7.5

In English service, also called butler service, all the food is placed on the table and the guest host serves the food.

The meat or main entree is placed directly in front of the host with warm plates stacked behind the main entree. The host serves a portion of the meat or main entree on a plate and passes it to the right so guests can pass it down to the hostess. If the main item is a roast or some item that has to be carved, the host carves it progressively, placing carved portions on plates for the passing to the hostess. The hostess, upon receiving the plate, adds the vegetables and other foods and then passes it to guests who pass it down until it reaches the proper guest. Sometimes, instead of this passing by guests, service personnel take the plates from the host or hostess and carries them to the proper guest. The host gets the last plate. Sometimes the hostess dishes the first plate and sets it down next to her; this is her plate since she is supposed to get the coldest food. No one eats until the hostess lifts her fork and touches the food on her plate.

The table is usually set according to the standard procedure of knife and spoons on the right and forks on the left. A bread knife will be on the bread and butter plate at a right angle to the forks. Dessert flatware will be placed typically above the service plate, if one is used. Desserts are usually plated by the hostess after being brought to her with proper serving dishes by the server. Wine glasses are placed to the left of the water glass or goblet. Coffee cups and saucers will also be brought to the hostess at the proper time and then she will serve the coffee. Cream and sugar are usually passed so guests may take their own.

Small tables or stands may be set at the right and left of the hostess and host so dishes on which food was served can be removed. However, servers may remove these and place them on a table or buffet in the dining area. Second helpings may be offered, so the foods will have to be available for this purpose, if it is done.

Chinese Service

The setting for Chinese service is quite traditional. It originated with Chinese families and was formalized for special occasions. Japanese and other Asian cultures have slightly different settings than the service described here, based on Chinese service. The setting for an elaborate meal is described below. However, since many meals may be much simpler, some of what is described will not be used at every meal.

There are many customs that must be followed in order not to offend guests. Thus, to not follow the traditional seating or to not serve the proper number of courses or the right foods can be taken as a subtle slight, purposely given.

Chinese service originated with families and was formalized for special occasions. The amount of labor used is minimal, but the amount of ware used is quite extensive since foods are usually served in individual dishes. Chopsticks are used for almost all foods, except for soup and dessert spoons. Food is dished into large platters, bowls, and other containers and brought to the table, where guests select their own food. Often there is a large selection of different foods. Many Chinese food operations set a circular wheel in the table center so guests can rotate it and share food.

Normally, for elaborate meals, a small plate about six inches wide is placed in the center cover. A rice bowl holding about a cup of rice is set to the right. A small cup for tea is slightly above the rice bowl or slightly to the left above the center plate. Chopsticks are on the right and often rest on a small stand. In some settings a long-handled, round spoon and bowl may be to the left of the chopsticks. Small dishes for sauces and condiments are above the small cover plate. A soup cup with a porcelain soup spoon may sit on the center plate or just above it away from the guest or to the left. (Soup is served at different times in Chinese meals; several may be served.) The napkin is folded and placed on the cover. A small wine cup may be on the table and set up among the small condiment dishes.

Food is brought to the table on platters or serving dishes and placed in the center of the table. Guests then help themselves to what they want. The dishes are usually elaborately garnished. It is not unusual to see sweet foods served at the beginning of and during the meal. Rice or soup is often served at the end of the meal.

The seating of guests is traditional. Large round tables are used at which as many as twelve people may be seated. There is an old legend in which an emperor was invited out for dinner. The host and hostess sat opposite the emperor with their backs to the door. They did this because they knew the emperor was in danger and enemies were out to kill him. In this manner the emperor could see who entered and be prepared to defend himself. The enemies did enter, but because of the seating arrangement and the help of those in the room, the emperor was able to escape injury. Even today the host and hostess sit with their backs to the door and guests sit opposite.

Usually no tablecloths are used, although napkins are. It is considered bad luck for a guest to be seated with the grain of the wood pointing toward him or her. It must run the opposite way, so the guest can face the length of the grain. Young people and individuals of lesser rank are grouped around the host and hostess. The next honored individual is seated to the right of the most honored guest.

As guests finish eating a dish, especially a soup, the servers should remove the empty dish and replace it with a fresh one. Wine cups, however, are not removed. Some guests may find it difficult to eat with chopsticks, so servers should be prepared to bring forks and other utensils. Knives will not be needed because the food is usually cut into bite-size pieces.

CHAPTER SUMMARY

French service is also called tableside service because it occurs there, with a chef de rang preparing items on a guéridon. A réchaud is used to heat foods. Foods are brought for preparation from the kitchen by the commis in a partially or non-prepared state so the tableside preparation is to finish the food. The commis also receives the food from the chef du rang and serves the guests. The chef de rang often takes beverage orders and serves them. A sommelier will handle wine ordering and service. A commis debarrasseur may be used to assist in the service.

French service is costly because of the amount of serving labor used, because it takes more space, and because the turnover rate is slowed down, because it takes longer. A large amount of equipment and ware must be used, and this increases cost. Because it is costly, the price charged for meals is high. It is, however, an elegant service, having much style and showmanship.

A chef de trancheur is often used to carve meats and poultry, working from a special voiture.

In Russian service, food is plated and garnished in the kitchen and brought to the table on platters or in serving dishes. Portions from platters are lifted and placed on the plates of guests. Russian service is slow, reducing turnover. It also uses more equipment than American service. Russian service has a great deal of elegance to it, but does not require the amount of labor, space, and equipment that French service does. This type of service is generally used in banquet and large catered events.

In American service, food is plated and garnished in the kitchen onto the dishes to be placed at the guest's cover. It is fast, uses a minimum of labor and equipment, saves space, and can be varied to suit the needs for serving breakfast, lunch, or dinner. It can also be varied from quite formal to quite casual. It is the most used kind of service.

English service is also called butler. Food is brought to the table in serving dishes, with vegetables and salads going to the hostess, and the meat or main dish going to the host. If carving is to be done, the host does it. The main dish is placed onto the dinner plate and passed down to the hostess, who adds the potato or starch item, vegetables, and any other food to be put onto the plate. If a sauce is served with the main entree, the host serves it. Plates are usually passed to and from the host and hostess by guests, but servers may do it. The service is gracious and intimate and lends itself to parties where people know each other well. It is also used in some clubs, but it is not common.

Chinese service is being seen more often. It is quite traditional and requires a fairly large number of separate dishes to complete a meal. Food is brought to the table on large serving dishes, often elaborately garnished, and guests serve themselves. Often a circular wheel is in the center of the table from which guests can get the foods they wish. The main eating utensils are chopsticks, but a metal or porcelain soup spoon will also be used.

KEY TERMS

chef de rang
chef de trancheur
commis debarasseur
commis de rang
crumbing
flambéing
le grand couvert
sommelier

CHAPTER REVIEW

1. What are the five classic styles used to serve food and beverages?
2. What are the characteristics of each that distinguish it from other forms of service?
3. What are the advantages and disadvantages of each?
4. Match each French service term on the left with its definition on the right.

_____(1) Chef de trancheur	a. A cart on which food is prepared at tableside
_____(2) Chef de rang	b. To ignite
_____(3) Guéridon	c. Carves meat
_____(4) Réchaud	d. Prepares food at tableside
_____(5) Sommelier	e. Used to heat food
_____(6) Flambé	f. Wine steward
_____(7) Captain	g. Head of service

5. Describe the two ways in which soup is served to guests using Russian service.
6. In which type of service would the following be used? The server places a spoon under the portion of food, and a fork over it, and then lifts the portion and places it on the guest's plate.
7. A porcelain soup spoon is used in what type of service?
8. In which type of service would the following occur? Coffee cups and saucers are on the table at the start of the meal.
9. What service descended from le grand couvert?
10. When deboning a fish, what should you remove first?
11. Match each phrase on the left with the type of service on the right with which it is associated. Letters will be used more than once.

_____ 1. This service style is also called butler service.	a. French service
_____ 2. Food is prepared in the kitchen and served to guests by servers from platters and bowls.	b. Russian service
_____ 3. Food is dished onto plates in the kitchen before being served to guests.	c. American service
_____ 4. Salad mixing and meat carving are performed tableside.	d. English service
_____ 5. This style gives chefs and cooks the most opportunity to garnish and plate food attractively and creatively.	

_____ 6. A team of servers, including a head server and various assistants, serves guests.

_____ 7. In this service style, it is appropriate to place ashtrays, sugar, cream, salt, pepper, and other condiments in the center of the table before guests are seated.

_____ 8. Many foods are served by the server using two spoons.

_____ 9. Food is sometimes covered with alcohol and set ablaze in front of the guests.

_____ 10. The server rests the serving platter on the arm or on the open palm.

a. French service

b. Russian service

c. American service

d. English service

Case Study

Setting Up a Service Staff

A new hotel is opening up in Las Vegas on the Strip. It will have five restaurants. The restaurant types include family style, steakhouse, seafood, Italian, and Chinese. In addition to these restaurants, the planners want an upscale restaurant with French service for high rollers. The manager of the hotel plans to have an average dinner check in the French service restaurant of more than $100 per person. It will be open only for dinners, seven days a week.

The new hotel's manager contacts a maitre d' hotel in another Las Vegas hotel who is known for his ability to organize and run a high-class French service, and the manager invites this person to be the manager of his hotel's French service restaurant. The maitre d' accepts and is able to bring with him the chef of the hotel's restaurant where he is working.

Thus, the maitre d' comes to the hotel with a chef he knows he can work with. He doesn't bother to hire a kitchen staff, leaving this task up to the chef, but he has to set up the French service in the front of the house himself.

Set up this French service staff. Plan for the operation to serve slightly more than 100 dinners a night.

SERVING THE MEAL

OUTLINE

I. Introduction
II. Steps in Serving
III. Greeting and Seating the Guests
  A. The First Approach
  B. The Introduction
  C. Presenting Menus
IV. General Rules and Procedures for Serving
  A. Serving Water and Ice
  B. Carrying Trays
  C. Handling China and Glassware
  D. Handling Flatware
V. Taking the Order
  A. Suggestive Selling
  B. Placing and Picking Up Orders
VI. Serving the Guests
  A. Breakfast Service
  B. Lunch Service
  C. Formal Dinner Service
  D. Booth Service
VII. Clearing Tables
VIII. Presenting the Check and Saying Goodbye
IX. Closing
X. Formal Dining
XI. The Busperson's Role
XII. Chapter Summary
XIII. Chapter Review
XIV. Case Studies

LEARNING OBJECTIVES

After reading this chapter, you should be able to:

- Oversee proper setting of tables, proper meal service, and clearing.
- Describe receiving correct payment from customers based on accurate guest checks.

Introduction

Every profession has its rules and procedures for accomplishing required tasks; table service is no exception. Up to this point, many general service rules and procedures have been given. This chapter focuses on the tasks of serving guests. First, we will cover some of the more general rules and techniques of handling trays and other service equipment. Next, rules and procedures for casual dining are covered, followed by special rules and procedures for formal dining.

Steps in Serving

Serving food and beverages involves a sequence of five steps: (1) greeting and seating guests, (2) taking the order, (3) serving, (4) clearing, and (5) presenting the check and saying goodbye. The tasks of each step vary according to different meals and type of operation. Thus, breakfast is served differently from dinner; a drive-in serves differently from a cafeteria; and counter service differs from table service. Service requirements also depend on whether guests want a leisurely or a quick meal. At breakfast, guests are usually in a hurry and things are done so that guests can be on their way. Lunches can be hurried or leisurely, and it is crucial for servers to take their cues from guests. At dinner, the pace is likely to be more leisurely. Banquet service, buffet service, and other specialized services have special requirements.

Greeting and Seating the Guests

The first employee that a customer comes in contact with represents the first opportunity to make your customer's experience a positive one. All employees, the owner, manager, host, and server, should know how to properly welcome a guest to the operation. How a customer is welcomed is dictated by the type of service your customer expects. Guests are typically greeted by a host or hostess or even the owner, but in other casual dining situations the greeter might be the server. The **greeter** should see that the greeting includes the most convivial elements: a pleasant attitude, a warm smile, eye contact, and a brief but welcoming phrase. If the guest has a reservation, it should be honored by immediate seating. However, this is not always possible and the guests might have to wait a few minutes. The procedure for handling this is discussed in Chapter 10 under reservations.

Hosts and hostesses must be well trained in the operation's procedures for seating people. With regular clientele, their desires for a table may be known. Some like to be seated where they are seen. Others do not. Many restaurants have smoking and nonsmoking sections, and guest preference should always be followed in this instance.

Hosts and hostesses should be alert and accommodating to guests' seating preferences. It is customary for hosts or servers to pull out the guest's chair so guests can seat themselves. Also be aware of special needs, such as those for people with disabilities and families with children.

The First Approach

The server's first contact with guests is crucial. At this point, guests make a summary of what to expect in service, and this often sets the size of the tip. The greeting, the seating, and the first approach to guests at the table should create a positive impression on the part of the guests and establish their estimate of the server's competence and ability to serve. The more pleasant one can make these first few minutes, the more likely it is that the server will have an easier and more pleasant time of serving, and that guests will enjoy their experience.

A friendly attitude is essential when dealing with guests, but the attitude should not be too familiar; some dignity should be observed. Do not indulge in unnecessary conversation or encourage familiarity.

The server or the **captain** (leader of a group of servers), if there is one, should help guests with their coats or other items. In some operations, the server is not expected to seat guests, but in some fine-dining establishments, the captain and server might do this.

Servers should make sure that their stations are ready to receive guests. Items such as candles, flowers, table tents, and place settings should be in place. The candle may or may not be lit, according to the operation. Some operations feel that unlit candles and upside-down glassware on the table indicates the table is not ready for guests, so it has glassware right side up when guests arrive. Flatware may or may not be on the table; in some operations, it might be placed only after orders are taken so the server knows what is required. In other operations, only a basic setting of flatware—knife, fork, and spoon—is set and the other items are added as needed. Unless an all-purpose wine glass is used, the proper wine glasses and other required glassware will be set after the wine order is taken. In full-service establishments, salt and pepper shakers may or may not be on the table. Condiments are not placed on the table until guests order. However, coffee shops and other more casual dining units often have salt and pepper shakers and condiments on the table when guests are seated.

Servers should be watchful of guests coming to their station. Let guests know you have seen them. Take a step forward to greet them and say the appropriate, "Good morning," "Good afternoon," or "Good evening" with a smile. If you know guests' names, use them. Take them to their table if the host or hostess does not do so, and ask if the table is suitable. At times this cannot be done because the server is busy with others. However, enough time should be taken to let the new guests know the server knows they are there. A short, "I'm sorry, we're so busy. I'll be with you in a minute," can help to give the server a chance to finish what is being done and come back to the guests.

The Introduction

In some operations, when everyone is seated, the server introduces himself or herself by name, saying something like, "My name is _____. I will be your server this evening." Some operations do not like servers giving their names and only have the server greet the guest. If there are complementary snacks, they should be brought immediately, and water should be poured. It is also appropriate at this time to pre-bus or remove extra place settings from a table. If a table is set for six, but there are only four customers, the server should ask if additional guests are expected. If no other guests will be joining the party, remove the two extra settings. This gives the customers more room on the table, and saves the operation money.

Presenting Menus

Menus are frequently handed out right away, but some operations prefer to wait until the premeal beverage order is taken. If a premeal beverage menu is given out, the regular menu is then given out later. The person handing out menus depends on the operation. In most operations, the manager, or host, gives them out after the guests have been seated. Menus might already be placed at each cover. Or the server gives them out after the server's introduction or after the beverage order has been taken. The server should hand out menus to each guest's left, unless space does not permit this.

It is traditional to hand guests their menus, either opened or unopened. It is discourteous to drop menus on the table so guests have to pick them up and open them. An operation might or might not specify that menus be given first to women. The server also might hand out a wine list, or a sommelier will do this separately.

After the menus have been given out, the server should describe all food and beverage specials along with their prices. Servers should have tasted all specials so they describe them honestly and appetizingly. The server should also mention items on the menu that are especially good or unique, or that management wants to push. A remark such as, "The chef is particularly proud of this dish. He is planning on using the recipe for the next culinary contest," might arouse interest in an item. A description of the basic methods of preparation of some dishes also arouses interest.

FIGURE 8.1

Servers can add a special touch to their service by explaining the menu to guests.

Servers should be familiar with all menu items, their ingredients, and the methods used to prepare them. It is embarrassing, and bad for business, to not be able to answer guests' questions about the menu. Guests also may wish to receive information about the nutritional qualities of some dishes. However, servers should not give out information unless they are positive they know the correct answer. If a customer needs to know if the specific ingredients in a recipe for medical or dietary reasons, the server should ask the kitchen and give the customer the correct answer.

General Rules and Procedures for Serving

Proper serving is a craft that, when done correctly, flows so smoothly it appears simple to the untrained eye, yet when the novice attempts to do the tasks required, they become a challenge to one's knowledge and serving skills. What seems so simple when observed becomes extremely difficult when it has to be done. Just organizing the task alone so the service proceeds in a logical manner becomes a trial. However, with sincere application in learning the basics of service and by acquiring some dexterity in handling trays, china, glassware, and flatware, the novice server can become quite proficient in serving patrons, and in a short time can handle quite competently a fairly good-sized station.

Servers cannot be considered professionals until they become well acquainted with handling all service equipment confidently. Although the restaurant industry is in constant evolution, the following discussion covers some of the recommended serving procedures and rules that apply to most forms of service. Proficiency in these procedures along with some others can go far in making a professional of the novice.

Some servers like to develop their own methods; some of these methods may be acceptable, but often they are not. The rules and procedures cited here have been tested over time and have been found to give the most satisfactory service.

As indicated throughout the various chapters in this book, there are many personal requirements of a server. The most basic one is the ability to quickly get food and drinks from order, to the kitchen, and these back the guests. This is a demanding task.

It is essential that servers observe what is right and are not allowed to develop the wrong habits from the beginning. We are all creatures of habit and once we adopt certain methods and become accustomed to them, it is found that one learns to work smoothly, easily, efficiently, and quickly, giving a desirable level of service.

All dining operations should establish a flow pattern that servers should follow when moving in the dining area and kitchen. Breaking the flow can cause accidents because a fellow worker may not expect someone to act in a varying manner. Management should plan the flow pattern to follow, and train the service staff to follow it.

Serving Water and Ice

One of the first things done is to give guests a glass of water with ice. When pouring at the table, the pitcher should be positioned two to three inches away from the glass rim. If it is too close there is a risk of touching and chipping the glass; if it is too far, even the most skillful server runs the risk of spilling. The glass should not be filled to the rim; two-thirds to three-fourths full is sufficient. Patrons dislike to handle a glass that is completely full. The busperson or server should be alert during the meal to see that water glasses are kept filled. Some operations provide a pitcher or carafe of water so guests may fill their own glasses.

Dining area supervisors should see that employees know that spills require *immediate* attention. If the spill is in a busy area, an employee should remain there and direct traffic around the spill. The employee should warn nearby customers and fellow employees of the spill. While clean-up is in progress, the employee should post a sign, such as "Caution—Wet Floor." The sign must be left in place until the area is safe. If the spill is liquid and cannot be cleaned up for a time, an approved absorbent compound may be used to contain the liquid. If water or chemicals are used in the clean-up process, the employee should avoid wetting any more area than necessary.

Carrying Trays

Carrying a tray is one of the first tasks a server should learn. Trays come in various shapes and sizes, but the most common are the 27-inch to 30-inch oval tray, used for large loads, and the 15-inch cocktail tray, used primarily to serve beverages. Trays are customarily carried by the left hand raised slightly above the left shoulder. For sanitary reasons, the edge of the tray should be at least four inches away from the head or neck. Servers' hair must not in any way come in contact with items.

Almost all lifting in foodservice establishments involves carrying, especially for servers who must carry trays or plates. Servers and buspersons need to plan a route so they will not bump into other employees and guests.

When servers and buspersons plan their route, they should check the condition of the floor along the route, and look closely for any hazards, such as pieces of furniture or equipment out of place, spills, sharp corners, narrow passageways, and stairs.

Employees should use their whole hand to grip the load, not just their fingers. They need to keep the load close to their body and centered. Other carrying procedures include the following:

- Keep the ears, shoulders, and hips going in the same direction.
- Keep elbows against sides, for additional balance and so nothing is bumped.
- Keep stomach muscles firm and the lower back tucked in. The load should be carried by the legs and hips, not the back.
- To turn, move the whole body as a unit instead of twisting at the waist. Face the load when lifting it and setting it down. This might feel somewhat robotic, but the back will benefit.

It is not recommended to carry trays on the tips of the fingers. Using the palm of the hand gives more firm support and better control. (An exception might be when one is carrying a tray through a crowded area; raising the tray on the fingertips may help raise the tray high enough to get through.) The cocktail tray can be carried securely between the thumb and index finger, with added help from the other fingers, giving better control. This is especially true when carrying such a tray with glasses filled for a party of six or more. It is also recommended when carrying tall, fragile stemware or large items such as coffee pots.

FIGURE 8.2
Trays should be carried on the palm of the hand to keep them stable and reduce the risk of dropping the food. Courtesy PhotoDisc, Inc.

Items on a smooth surface of a tray can slide; trays should be covered with cork, plastic, or a rubber mat to prevent this. A wet napkin spread over the surface of a smooth tray can also help reduce sliding. All trays should be kept clean for use.

Trays should not be placed on a table being cleared. Instead, the tray should be placed on a **tray jack** (also called a tray stand) near the table being cleared. (A bus cart can replace the tray on the tray stand.) Trays should never be placed on a table at which guests are seated. They can be placed on a table, primarily in a large banquet or buffet operation, once guests have gone.

Some servers, especially in setting tables, like to carry stemware without trays by inserting the glasses by their stems through the fingers. While this is efficient and speeds table setting, for reasons of safety the use of a tray is recommended.

Perfect balance is the secret to carrying trays safely. Balance must be absolute to allow freedom of action and maneuverability when opening a door with the other hand, making turns, or just carrying the tray. Good balance can be obtained by distributing the weight of the items equally throughout the surface. The heaviest items should be placed in the tray's center with the lighter ones on the outside. When a tray is properly loaded by placing the palm of the hand exactly in the bottom center, there is no tilting from any side and the tray rests securely, with the fingers able to control any slight variation.

In some operations, tray jacks are permanently set up, ready to receive the loaded tray. In other establishments, the server brings the jack to the table with the free hand, opens it next to the table, and places the tray on it. If the tray is quite heavy, servers should place the jack prior to bringing out the tray. Sometimes another employee may come ahead of the server carrying the heavily loaded tray, open the jack, and help the employee place the tray on the stand. Once the tray is in place, service occurs with one or two dishes being removed from the tray at one time. When the tray is emptied, it is removed.

When resting a heavy tray on a tray jack, bend the knees slightly and put the tray down cautiously while holding it with both hands. Many times it is better to ask a fellow server to help place the tray on the jack.

Loading trays require attention, not only to achieve balance, but to get a maximum, but safe, load. This takes experience, and servers should note how more experienced servers load their trays. **Food covers** are used for many items. This allows another plate of similar size to be placed on top of the first plate. The cover also helps keep foods hot. It is possible using covers and stacking in this manner to get eight dishes on one tray. (In banquets, as many as sixteen entrees may be put on one tray.) Being able to stack in this manner gives good balance while at the same time helping the server to get more entrees to guests, saving travel time. It's also important to carefully load beverages so that the server can bring all the beverages to the guests in one trip, without spilling.

Servers should take care not to place hot and cold items on top of one another or even allow them to touch. Pots holding hot liquids should not be filled so full that the hot liquid easily spills from the spouts when the tray is carried. Most experienced servers also do not put cups on saucers or dishes of varying sizes or shapes on top of one another. Such items fail to nest and can slide easily. When carrying filled containers that will be set on another saucer, dish, or tray, place the filled container on the tray and not the item on which it will sit. In this way, any spills occur on the tray and not on the item on which the container will sit.

Handling China and Glassware

Sometimes a server is seen carrying a tall stack of dishes with one hand underneath while holding the stack firmly against the chest with the other hand. This is both risky and unsanitary. Keep the fingers off the rim of the dish. If carried as a stack, put both hands underneath and hold the stack away from the body. A **serviette** or service towel may be wrapped around the stack to help hold it in place. Do not try to carry too high a stack.

Handle tumblers from their bases and stemware from the stems. Handle dishes with the hand under and with the thumb along the rim of the plate. Watch to see that soiled, chipped, or cracked china or glassware is never used. Fruit juice and cocktail glasses, cereal dishes, soup bowls, dessert dishes, and pots should be set on underliners when placed on the table. Set iced beverages on an underliner or coaster. In some operations, a doily is set on the underliner before placing the item down.

If an item is hot while setting it before a guest, the server should use a fold-over-twice serviette or napkin between the thumb and index finger to protect the fingers. The server should also notify the guest that the item is very hot. If the guest does not hear the warning, the server should repeat the warning until the guest has acknowledged it.

Handling Flatware

Handling flatware properly is as important as handling china and glassware. Never touch the part of flatware that will go into a guest's mouth; hold only its handle. All flatware should be wiped thoroughly before setting it on the table, not only during mise en place but also at other times. If flatware is extremely spotted, management should be notified so it can check dishwashing procedures. In particular, the blades of knives need to be wiped more than once; the spots from hard water and films left by cleaning agents and dishwashing machines are visible on the blades more than any other place. A cloth or towel wet with a bit of vinegar and water is good for cleaning; the scent of vinegar quickly disappears but the acid of the vinegar removes the alkaline agents that spot the object. Do not put bent, tarnished, or soiled flatware at a cover. Placing flatware properly on the table is an essential of good service.

When resetting tables, some servers prefer to use a cocktail tray loaded with flatware organized so like ware is together. Others prefer to use flatware caddies loaded by kind in separate compartments. Whenever carrying flatware around the dining room, at least a plate with a napkin on it should be used. In handling spoons, forks, and knives, the fingers should touch only the lower part of the handle and never the business end.

Taking the Order

After the premeal beverages have been served and there are no reorders, the server should ask if the guests have made their selections. It is not unusual for some to have not even looked at the menu; the server should suggest appetizers and offer to return a little later. An alert server will note when the guests have seemingly made their choices, or when they may need assistance in making selections. The server should be there promptly to take care of guests' needs. Readiness to order is often indicated when a guest closes the menu and sets it down.

FIGURE 8.3

When guests are ready to order, the server should be prompt and friendly, and ready to make suggestions when questions are asked. Courtesy PhotoDisc, Inc.

Guests may have difficulty in finding items or in reading the menu. Servers should note this and give assistance. Sometimes the guest cannot read well. The server may, in this instance, read the menu to the guest, explaining each item. Some may wish to know how items are prepared. Guests often want to know about a food's type, quality, grade, or preparation. For example, "Are the strawberries on the shortcake fresh or frozen?" "Are the oysters bluepoints or pacifics?" "Is the beef Choice or Prime?" "Are your pies baked here?" The time it takes to prepare might be important to some guests, and servers should be prepared to answer on any time questions. Servers must be prepared to answer a host of questions.

Some guests know immediately what they want; others may be undecided because they are not familiar with the service, are not hungry, have limited funds, or do not understand the menu. Servers can be a big help with these guests.

In most cases, the server takes the order, but in the more formal situations a captain might. Orders are often written on a **guest check.** These are usually serially numbered; servers are given a certain number of such checks, which they must sign for, before a shift. Some operations hold servers responsible for checks that disappear. There may be a standard charge levied against a server if a check is missing, which can occur with someone walking out without paying. Unused checks at shift's end are returned and credited to the server. However, where the POS system is used, as it is in many operations, servers are no longer held responsible.

When taking the order, the server should ask whether the guests want separate checks or one check. The server should stand at the left of the guest ordering. For small groups, the server may stand in one place where each guest can be seen and heard. Do not hover over guests. Be sure to get complete information, such as the kind of vegetable, the doneness of the meat, and so on. If an item takes long to prepare, tell the guest. It is desirable that servers also try to learn who is to pay the check so that presenting the check is clear.

Servers use systems on their order pads to help them remember who ordered what. The most common method is to establish the cover directly pointing toward a set point in the dining room as cover 1 or A, and to number the others clockwise. This standard system of guest order location is often called the **pivot system.** Cover No. 1 starts with the guest always sitting with his or her back to the kitchen. When the server returns with the orders, the server knows that this cover gets the New York steak cooked medium rare with a baked potato, house salad, and green beans. Cover No. 2 gets the chicken fricassee, mashed potatoes, peas, and tomato salad. No one is sitting at cover No. 3, but cover No. 4 gets the seafood salad and toast. Beverage and dessert orders will follow; the server will code these properly so service is easy and fast. All servers should follow the same system. In this way other servers can step in and work the table without asking a lot of questions.

Today much order taking is computerized. Servers send orders through the computer in the dining room to cooks in specific kitchen stations. It is possible for servers using this device to code in cover locations so that the server knows which guest ordered what when the orders are brought to the table. Usually the time the order is placed, the time it is ready, and the time it is picked up are recorded so management has a check on how long the flow of order taking to service takes and where to spot responsibility for delays. The computer system will also print sales tickets. Handheld computer systems are increasing in popularity, but, they still require an operational system, like a pivot system, for taking orders.

In any **guest-check system,** servers should write out orders in a legible and organized manner. If second and third copies are made, the server should press hard so each copy is readable. Good copies and legibility can help management in making check-duplicate reconciliation or any other control system used. For clearer writing, a booklet or menu under the pad or check can be used. When writing on a plain pad, it is recommended that a line be drawn after each course. This indicates to cooks when items are desired in the meal. (There might be a request where the guest asks for the salad to be served before the soup, or the salad to be served with the entrees, etc.)

Abbreviations are necessary to save time. (See **Exhibit 8.1.**) However, they should be standardized. Every server should know and use the abbreviations used in the operation. Following are some common abbreviations:

Extra rare (blue)	XR	Very well done	VWD
Rare	R	Chicken sautéed	Ch saut
Medium rare	MR	Chicken fried	Ch f
Medium	M	Steak	St
Medium well	MW	Roast Beef	Rst B
Well done	WD	Prime Rib	PR
Baked	BKD	Sour cream	SC
Creamed	CR		

It is helpful to underline extreme degrees of doneness, such as extra rare and very well done. This reinforces the fact that the guest wants the item that way. The server should also note any special requests. At the beginning of the shift, abbreviations for specials will be given by management to both servers and kitchen staff. Abbreviations help speed up order taking. If they later cause confusion and errors, then their purpose is lost and it would be far better not to use them.

Always repeat the order after the guest gives it. This prevents errors and misunderstandings. Be sure to get all the information needed, such as type of salad dressing, doneness of meat, and any special requests. Sometimes a guest may just give the entree order, close the menu, and hand it to the order taker. It is then necessary for the order taker to find out what the guest desires for the rest of the meal, such as an appetizer, the kind of soup or salad desired, beverage, and perhaps even the dessert, although dessert orders are usually taken after everyone has finished with the entree.

EXHIBIT 8.1 Guest Check with Abbreviations

Using standard abbreviations saves time when taking orders.

GUEST CHECK

TABLE NO.	NO. PERSONS	CHECK NO.	SERVER NO.
		339540	

St, MR
PR, R
2 bkd pot, sc on side
2 cr spin

TAX

Suggestive Selling

Servers do more than take orders, serve, and clear dishes. They are important sales people, and a great part of the sales function is suggestive selling. More than a sales tool, suggestive selling helps guests make up their minds so they are pleased with what they have selected. It is an essential part of excellent service.

Every encounter with a guest is different. You will have to ask questions and watch for cues to find out what guests like and dislike, how much time they have, and how much they want to spend. Remember that guests can neither see nor taste menu items before ordering them. They depend on servers to help them make the right decision.

Appropriate attitude, dress, and confidence will encourage guests to take your suggestions. Let guests know that you're there to please them, and that they can trust you. It is important not to oversell; don't be pushy. This brings resentment and distrust. But don't be afraid to suggest; the bigger the check, the bigger the tip.

When taking orders, servers should use open-ended rather than closed-ended questions. Open-ended questions allow further discussion of the subject, while closed-ended questions stop it. By asking an open-ended question, the server has a chance to discover something a guest might like. An open-ended question might be, "What would you like for dessert?" Even if the answer is "Nothing. I don't eat dessert," the server has a chance to make a sale by suggesting other options, such as a low-calorie fresh fruit plate or a cordial or other nondessert item. A closed-ended question might be, "Do you want an order of sautéed fresh mushrooms to go with your steak?" In this instance, the only response is yes or no.

Here are some tips for effective suggestive selling:

- Suggest beverages and appetizers to start a meal.
- Suggest premium liquors when guests order generic drinks.
- Suggest fresh fruit if the guest hesitates on the desserts.
- Suggest side orders with entrees.
- Suggest definite menu items; don't ask, "Will there be anything else?"
- Know the menu and suggest low-calorie items when appropriate.
- If the order will take some time to prepare, suggest an appetizer.
- If the guest orders an a la carte item and it is also on the dinner menu, suggest the complete dinner.
- Use appetizing words, such as *steaming, sweet, spicy, juicy, fresh, savory,* and *refreshing.*
- Sometimes certain items are also sold for takeout, such as pies and cakes, salad dressings, and some prepared foods. If a guest particularly likes an item, an alert server recommends that the guest purchase the item to take home (for a bigger check).
- Suggest desserts, desserts to split, and after-dinner drinks.

Placing and Picking Up Orders

When orders are sent by computer to the kitchen, servers do not have to go there to see that the right cooks get their orders. Servers are given a key, code number, or authorizing card to enter the computer system. After entering, the server usually has an identifying number or code. The table number, number of guests, and the check number is usually entered after this is done. The time may be automatically recorded. The guest check is now inserted into the computer and the orders are entered. Often there is a **preset keyboard** or **touch screen** with almost all items on the menu listed. All the server has to do is touch one of these keys to order items. In some cases, the machine may ask for further information such as doneness of meat, or flavor of ice cream, and the server must then add this information. When all orders are placed, the server punches a key to print out the guest check.

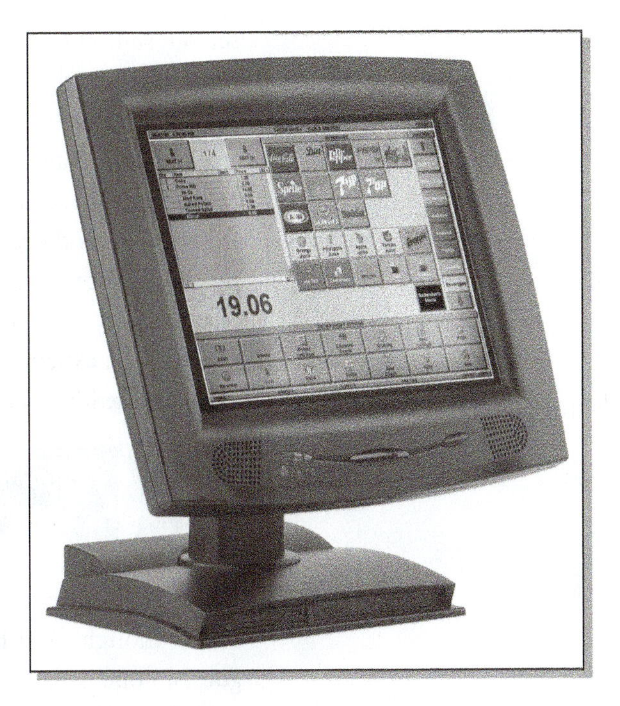

FIGURE 8.4
Many establishments today have automated systems allowing the server to keep track of a table's order and print out a check at the end of the service. Courtesy Action Systems, Inc.

With manual systems, servers must still place their orders in the kitchen. In some fine dining operations, the order in the kitchen is first checked by a checker, who then places the order. In formal dining, an **aboyeur,** or expediter, may take the order and place it with the proper sections. In some kitchens, servers call out their orders to the proper preparation personnel, but many avoid this because it can cause confusion and noise in the kitchen when several or more servers are calling out their orders. Quieter methods are often used. In the **spindle method,** servers put the order on a spindle for cooks to remove. In some operations, orders are placed on a rotating wheel, which keeps them in order as placed. The cooks then rotate the wheel to arrive at orders in sequence. In larger kitchens, servers must rewrite parts of the order so these separate parts can be placed in the proper section for preparation. Thus, a cold plate order might be separately written and go to the cold food section, the roast beef order rewritten to go to the steam table section, and a broiled steak order rewritten for the broiler section.

Various methods are used to notify servers that orders are ready. It is possible to have pagers send a signal to a specific server. Lights can be set in a dining area, and when this light is on, the server knows the kitchen is signaling that the order is ready. One novel way is to have a large clock that can light up an assigned number of a server.

Before taking an order from the kitchen, the server should check to see that everything ordered is there and is correct. (In some cases, a checker does this.) The server should take something out that does not appear right. Dished-up food should be neatly placed; garnishes should be right and attractive. Food spills on dish rims should be wiped away. Do this with a serviette, or towel, in the kitchen and not in front of guests. Servers should use a serviette or napkin to pick up hot dishes.

Servers should not pick up a course of a meal and bring it into the dining area until the previous course has been finished and cleared. However, in some faster-service operations, it is acceptable to clear one course while serving the next.

Servers need to organize their orders on pickup so everything is on hand when service at the table begins. One must be sure in the kitchen to pick up the correct items. Different menu items can look very similar, and if in doubt servers should ask the cook which meal they ordered. In some cases, cold foods should be brought to the service station before the hot ones, to be ready for service. When the hot items are brought from the kitchen, the cold and hot items are then served together. Pickup may require getting order items from various kitchen sections. For example, hot entrees may be waiting on a counter under a heat lamp, hot rolls may be waiting in a roll or bread warmer, and garnishes and salads may be in a refrigerator. Pick up hot foods last. Good organization simplifies the service task at the table. When the pickup is complete, a last-minute check should be made to see that the food is the correct temperature, and has an overall pleasing appearance and superior quality. Some operations use expediters who do the picking up, and they act as the link between service and the kitchen.

Serving the Guests

Serving the order is the total of all efforts of everyone involved in the operation. Much thought and labor go into the production of the items servers put before guests; none of this should be lost during service. As noted in Chapter 4, everything must be ready and in its place. All mise en place must be done before bringing food to guests. All supplementary serviceware must be at the server station.

Where specific food items are placed depends on the table setting used. First courses, soup, and appetizers of single servings are placed directly in front of the guest. In placing the entree, see that the main food item is in front of the guest. Appetizers to be shared are placed in the center of the table with appetizer plates placed before guests. Entrees are also placed directly in front of guests. Other items, such as bread and salads, are placed to the right or left of the guest. Side salads are placed to the left, while breads are placed to the right.

In more formal meals, the placement of certain items are more or less prescribed. Beverage glassware should be on the right. This text indicates these, but there is so much variation today, that one might say that there is no consistent standard. Follow the standards of the operation, and try to make things convenient for guests. Managers in setting standards should see they provide the type of service guests want and that servers are able to meet.

Servers, whenever possible, should use the left hand to place and remove dishes when serving at the guest's left, and the right hand when working at the guest's right. This allows the server to have free arm movement and avoid colliding with guests' arms. Never reach in front of guests, or reach across one guest to serve another. Present dishes from which guests serve themselves on the guest's left, holding the dish so the guest can conveniently help himself of herself. Set serving flatware on the right side of the dish with handles turned toward the guest so it is easy to pick up the item for self-service. In most operations, salt and pepper shakers, sugar bowls, and condiments are placed in the center of the table. In booths and on tables set against a wall, these items are placed on the wall side. Bread trays and baskets are placed in the table's center. Cups and saucers go to the right of the guest, with the cup handle to the right.

Normally, entrees are served with the main item placed on the lower part of the plate closest to the guest, called the six o'clock position on the plate. The garnish and whatever side items are on the plate should be neatly arranged to make an attractive presentation.

Breakfast Service

Breakfast service must usually be fast. Fruits and juices should be served chilled. Milk is required with cereals and some operations also offer a finer sugar than regular, called berry sugar. Toast should be freshly made, and buttered or not, as the guest indicates. Hot breads should be hot and fresh. Hot cakes and waffles should be served as soon as possible; they lose quality rapidly as they cool. Eggs cooked to order must be correct, and served immediately. Hot beverages should be very hot.

Many operations today offer buffet breakfasts. This helps guests move quickly through the meal and allows them to take what they wish. Some guests may not wish as much food as is offered on a buffet, so most operations find that in addition to the buffet breakfast, they still must offer a menu and allow guests to select what they want.

Lunch Service

A lunch may consist of only one dish, such as a soup or salad, and a beverage. In a luncheon with courses, each course is placed directly in front of the guest. Vegetable dishes are placed above the entree to the right. Salad is placed on the left, with bread to the left of the salad. If a chilled beverage is served, place it to the right and a bit below the water glass. The handles of cups on the saucer should be turned to the right at the three o'clock position. At formal luncheons, servers *crumb* the table between courses. (Crumbing is removing crumbs and other items from the table surface.)

Formal Dinner Service

The first course in a full-course meal is placed directly in front of the guest. If a cocktail fork or other utensil is needed, place it on the right side of the plate or to the right of the plate. If flatware is already in place for the rest of the meal, this appetizer utensil is placed to the right of these. If guests are to serve themselves from a dish, place appetizer plates in front of the guest to the left. Sometimes a salad is served as a first course instead of an appetizer. The placement is the same and, if served from a dish, service again is from the left.

In some cases a finger bowl may be served after the first course, especially after finger foods such as cracked crab. The water in the bowl should be warm and have a lemon slice floating in it. A new napkin should be offered.

Soup is typically served as a second course. The soup bowl is placed on a serving plate, and put directly in front of the guest. The soup spoon should be to the right. Offer crackers. Crumb between courses as needed.

Entrees are placed directly in front of the guest. A vegetable dish, if used, is placed above and to the right. Side salads are placed to the left of the forks. The butter plate is to the right above the knife. Sometimes guests serve themselves from a platter. Put an empty, warm dinner plate directly in front of the guest with the platter and serving utensils above this.

At the more formal dinners, the salad is served after the entree and is set directly in front of the guest after the entree has been removed. Sometimes a salad bowl is offered; if so, the salad plate goes directly in front of the guest with service from the left.

In some operations, a fresh fruit sorbet is automatically offered before the entree. It is usually served in a tulip champagne glass over a doily and underliner with a teaspoon. A flower petal or lemon wheel can be used to decorate the glass.

Often dessert utensils are not placed with the original table setting but are brought in with the dessert. The dessert is set directly in front of the guest. If the dessert is triangular in shape, such as a piece of pie, place the point toward the guest. Sometimes there may be some doubt as to whether a guest would like a spoon, fork, or some other utensil. In this case, place both at the guest's place and let the guest decide.

FIGURE 8.5
Boothlike service requires the server to lean over guests furthest from the wall. Courtesy PhotoDisc, Inc.

Booth Service

Booth service is often difficult because of the need to serve guests at the far end of the table. Service to guests seated there should come first to avoid having the server reach across dishes served to those closer to the server. Serving those at the far end of the table usually requires the server to lean over and reach to get the items in their proper place. To serve those on the right, place the left hip against the table and with the left hand, reach out and set the items down. For the guests on the left, place the right hip against the table and reach out with the right hand. In serving a small booth or a dish used by all, it is acceptable for the server to stand flat against the booth and reach over. Because booths often have less space than tables, it may be necessary to clear items as soon as possible.

Clearing Tables

In some operations clearing is not done until all guests have finished the course. In other operations it is acceptable to clear as each guest finishes. Some guests do not like to have dirty dishes in front of them and may ask for their removal. In such a case, the request is followed in spite of house rules. Servers should watch to see if guests wish their dishes to be cleared. Some signal this by setting their eating utensils down on the plate. If in doubt, the server should ask the guest if he or she has finished.

Clearing is done to the right using the right hand. Left-handed servers can make an exception to this rule if they don't feel comfortable handling certain items, such as heavy entree plates.

If beverages are not finished, leave them at the table unless the guests want them to be removed. Offer after-dinner drinks. Leave hot tea and coffee on the table for the dessert course. Water glasses remain on the table as long as the guest remains; keep them filled. Cracker wrappers and other miscellaneous items should also be cleared.

Crumb a table, with or without a tablecloth. If a crumber (or crumb brush) is not available, the blade of a dinner knife or a napkin folded into a roll shape will do the job. Do not sweep loose food particles onto the floor; use a plate or small tray covered with a napkin. The objective of crumbing is to get the table neat and clean and ready for dessert; coffee, and after-dinner drink service.

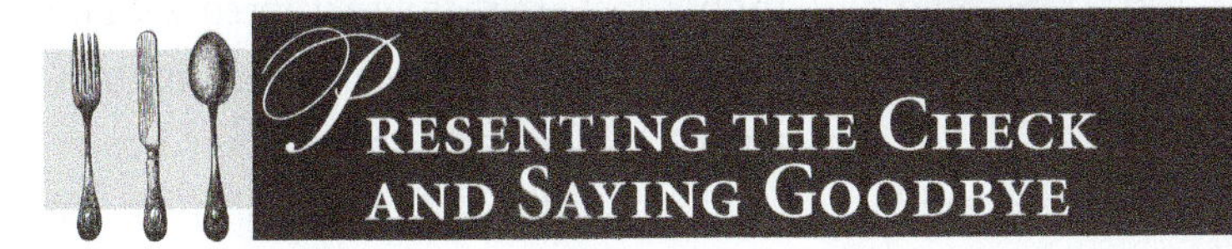

Presenting the Check and Saying Goodbye

When guests have finished and are ready to go, they resent having to wait for the check. The server, as well as the host or hostess, should be watching for signals from the guests that they are ready to leave, such as reaching for coats, purses, or packages. The last thing a server wants is a guest shouting, "Please bring the check!" Don't let guests rise and start to leave without the check, forcing you to rush over with the check. Difficulty getting the check can cause ill feelings, which might make guests decide not to come back or leave a paltry tip.

Present the check after the last clearing has occurred. It is an act of courtesy to ask, before presenting the check, if anything more is desired—even if the guest is seemingly in a hurry—so no oversight occurs.

The accurately totaled check may be presented in two ways. It may be brought to the table and placed face down on the table. This signals that the guest is to pay the cashier. It is appropriate to say something like, "Thank you. I have enjoyed serving you. Please come back again. You may pay the cashier."

In the second scenario, the server is expected to collect payment for the meal. The check is brought to the person paying the bill and is presented face down and then the server departs. The guest leaves payment plus a tip and leaves. If the guest does not have the proper amount, an amount greater than the check total is left and the server deposits the right amount with the cashier and brings back the change on the tray and leaves it with a "thank you." The guest picks up the change and normally leaves a tip. Servers can encourage a good tip by bringing guests' change in some single dollar bills instead of only larger bills.

FIGURE 8.6

Presenting the check is often the last impression that guests will have of their server. Courtesy Action Systems, Inc.

It is common to bring a receipt torn from the check for the guest. Servers should never manipulate the receipt amount to be larger than the actual bill, as this is a form of fraud.

Payment today is often made by cash or credit card. However, if payment is by personal check, the server usually must get approval for acceptance from an approved authority Always read checks to see that they are completed accurately. Most operations ask for identification from the guest.

Credit cards are much more common. Most are readily accepted with no identification because payment is guaranteed by the bank or credit card company. Always check the card for a valid date, and get authorization from the credit card company. The guest must sign the credit card receipt; the name and signature must match those on the card. Tips can be added to the receipt. The tip is then paid in cash to the server by management, which is reimbursed when the charge is paid.

It is unprofessional for a server to show in any way that a tip is expected. It is also unprofessional to show disappointment in the amount of the tip. Tips are given to show appreciation of good service. A tip lower than expected might be showing that the service was less than expected. Servers should know also that some guests are unfamiliar with American tip practices. In Europe, for example, tips (**gratuities**) are often included automatically in the check total. This is often true in the United States with larger groups.

Guest checks should be used for all sales transactions by all servers, bartenders, or others who make sales. Guest checks should be ordered only by the management, kept in a locked secure place, and issued only in recorded lots. Checks supplies should be frequently audited to ensure that none are missing.

Traditional cash register systems require the register operator to open the register to receive and dispense money. Enhanced point of sale systems reduce or eliminate this need, and may allow several order-entry terminals to feed orders to the kitchen, central register, and inventory system. When a POS computer is used, all payment is recorded automatically. If a payment is not recorded, the check remains active. Management can thus see from time to time which checks are still out and for how long. Walkouts can be discovered more quickly.

Many restaurants now use touch-screen **POS systems** (point-of-sale systems) that allow servers to close out their own checks and get them quickly to guests. These systems also process credit card purchases through computer-system links with national credit card companies.

Before guests leave, servers should be sure to thank them again and invite them to return. This lets them leave with a good feeling about the operation and an incentive to return.

Closing

When the serving period has ended, there are closing duties. Soiled linen needs to be counted, bagged, and placed in proper storage. All unused linen should be folded and stored with other clean linen. Unused clean flatware and other similar items should be returned to proper storage. Condiments should be wiped-clean and stored properly. Salt and pepper shakers, sugar bowls, and other similar items may be placed on trays to take to a work station to be replenished. Butter should be covered and stored under refrigeration. The work station should be placed in order. Tables needing fresh linen should receive it, and tables should be reset for the next meal, with clean glasses and flatware. (Some operations do not want overnight setting.)

There will be sidework that must be done. Usually management draws up a list of these tasks so servers are reminded about what needs to be done.

Closing after a serving period varies according to whether there is another meal period coming or it is the end of the day. It is important that those who close for the night see that those who come in to start the morning shift are not unduly hampered because the evening group did not properly prepare for the next meal.

Formal Dining

Formal dining is characterized by a number of factors, including:

- The dress code is usually more strict and formal.
- The food and service are usually quite elaborate.
- Menus are also quite elaborate.
- The decor is elegant.
- More servers are used.
- The atmosphere is reserved, quiet, and peaceful.
- The price will reflect the cost of these extras.

Generally speaking, the basic service mechanics mentioned for casual dining are also applicable in formal dining, but their execution is more elaborate. For example, butter pieces in casual dining are usually squares, while in formal dining they might be in the shapes of rosettes or flowery curls, requiring more delicate handling by servers.

When French service is used, much of the service is from the *guéridon.* American and Russian service are also used.

In formal service, wine is featured much more as a meal accompaniment; a sommelier is usually on staff. Wine service is also more elaborate and formalized. All servers in fine dining must be well trained, and work might be specific to service positions. Thus, a chef de rang's tasks revolve mostly around the table, while the commis du rang will leave the table to bring food from the kitchen and do most of the serving. Captains may do special work such as deboning a fish or preparing Crêpes Suzette.

The Busperson's Role

After guests have been seated and menus have been given out, often the next person to make contact with guests is the busperson. Buspersons might be young, but some like the work and stay in it for a lifetime. The position can be professional and lead to many job satisfactions. Often, besides a minimum wage, the busperson receives 15 percent of the food server's gratuity. In other food services, gratuities are pooled and distributed equally between all service personnel. Also, jobs often are not plentiful and the foodservice industry usually provides one. Some just like the work and stay in it.

The busperson's first responsibility is to provide water, butter, and perhaps bread after guests are seated and given menus. Bread baskets should be lined with a napkin because it is more sanitary, keeps the bread warm for a longer time, and looks appropriate and correct. Butter should be served cold but not on ice. (Once the butter softens and the ice cubes melt, the butter is apt to look messy and unappetizing.) Most often, water is poured at the table, but water in glasses can be brought to the table on a cocktail tray. (This can be heavy when serving a party of six or more. The procedure for handling a heavy tray is discussed earlier in this chapter.) There are many other duties required of buspersons, including the following:

- Bring in foods from the kitchen.
- See that ice and water are always at hand.
- Assist food server in removing soiled items. Always separate china, glassware, and flatware. Soiled items should be stacked according to shape and size to allow more room and give proper support to stacks. (This allows more to be stacked on the tray and saves on trips. It also helps the dishwasher.)
- Provide any supplementary service items needed by guests.
- Assist servers in beverage service.
- Refill coffee and tea orders.
- Ensure all condiments and service supplies, such as cream, sugar, lemon, teaspoons, cups, and saucers are readily available.
- Help maintain linen, equipment, and serviceware in an orderly manner in the various storage areas.
- Be available at all times for guests' special requests.
- After guests' departure, reset tables.

Here are some guidelines for doing the busperson's job effectively:

- Keep all equipment organized (a place for everything and everything in its place).
- Avoid overstacking items to guard against breakage and accidents.
- Always keep safety in mind; act immediately when there is broken china, glass, or spilled liquids on the floor.
- Walk, don't run. Perform tasks in a systematic manner.
- Try not to travel empty-handed. There is usually something to be carried in or out of the dining room or to work areas.
- Do not engage in long conversations with guests unless encouraged by guests.
- Good communication skills help make the job flow more smoothly and easily. Work to have open communication with other serving personnel and kitchen workers.
- At the end of a shift, leave the stations immaculate. Wipe down all condiment containers, bread warmers, cutting boards, bus pans, and carts, and make certain that all soiled linen is counted and properly bagged. Act immediately when you notice that the restaurant is running low on cleaning and paper supplies, equipment, or serviceware. Some operations like to have tables set up and prepared for the next shift. Others do not. Whether this is to be done or not is a management decision.

Throughout the course of the meal, the busperson should constantly remember one of the most crucial components of service—anticipating guests' needs. Service cannot be of good quality if guest needs are not anticipated. Patrons can become exasperated if during the meal they have to continuously ask for more water, bread, and butter. The novice has to focus attention on the table until, after a little practice and experience, it will become second nature to spot these needs, even from a distance. The busperson's approach should be immediate and courteous.

CHAPTER SUMMARY

Some general rules and procedures used in all types of operations include serving water and ice, carrying trays, handling china and glassware, and handling flatware.

Greeting and seating guests is the first step in good service. This task is very important since it gives a good impression to guests as to what is to follow. An important point in this first step in service is that servers should see that their station is ready to serve guests. This includes proper table setting, putting glasses upright ready for use, and so on. The seating of guests should not be a random function; those who seat guests need to make quick and intuitive guesses as to where to place guests.

Servers should greet guests with a smile and a welcome that lets the guests know that the server is happy to serve them. It is proper for the server to introduce himself or herself, hand out menus, and to help them make up their minds. Different operations present menus to guests differently. It is crucial that servers know the menu and are able to interpret it and describe items for guests. Often guests have questions or need help in making their selections, and the server is responsible for explaining what menu items are and how they are made. Servers should also be adept and comfortable with suggestive selling, not only to increase check totals and tips, but also to please guests.

Servers should use some system to write up orders so they know which guests placed which orders. Mechanized devices and computers transfer orders automatically to the kitchen and record sales. Order taking is made much easier and quicker when servers use abbreviations.

In noncomputerized operations, servers call out orders, place orders on a spindle or wheel, or give orders to an aboyeur or announcer.

The serving of the order is the culmination of all efforts of service and production. The main item of each course should be placed at the six o'clock position. Most items are served using the left hand on the guest's left, except beverages, which are served using the right hand at the guest's right.

Some of the essential rules and procedures for serving breakfast and lunch are covered. In dinner service, there often is a specific sequence of courses. Courses will be placed by servers or guests might serve themselves from serving dishes. Special considerations are discussed for booth service. Items should be cleared properly from tables and the table crumbed.

The server usually lays the check face down on the table near the person who is to pay it. Knowing who is to get the check is important because servers can cause some embarrassment if they do not know. If the server is to collect, the check is presented on a small tray or plate. The server receives payment and brings back any change along with a receipt. The guest then leaves the tip on the tray or plate. Credit card payment is common today. The server should always be sure to thank guests and invite them back.

Service does not end when guests leave. There is much to be done to be ready for the next service at that table. It must be reset, and the area made presentable. When closing out the meal, there are other things that must be done so as to be ready for the next meal to come. If the operation is to close for the night, other tasks must be done.

Formal service is characterized by how guests dress, the kind of place in which it occurs, special kinds of foods and menus, table preparation and service, and other factors. French service using the guéridon is common, but American and Russian service are seen. A few special rules and procedures for formal service are discussed.

The importance of the busperson in accomplishing service cannot be understated. Their tasks help make the meal flow smoothly. Servers often share tips with buspersons. The busperson's role can be summed up in the saying, "In every way, support the server."

Key Terms

aboyeur
captain
food covers
gratuities
greeter
guest check
guest-check system
pivot system
POS system
preset keyboard
serviette
spindle method
touch screen
tray jack

CHAPTER REVIEW

1. Where are heavy items placed when loading a tray? Where are lighter items placed?
2. What precaution must be taken when loading both hot and cold items onto a tray?
3. Why are the surfaces of trays often covered with cork? If a tray is not covered with cork, what can be done to give a similar effect?
4. Who normally first greets guests?
5. What is an appropriate way for a server to introduce herself or himself to guests at a table?
6. Why is it good for a server to help guests by suggestive selling? How can suggestive selling help the server?
7. If a guest orders a steak medium rare, how would you write it as an abbreviation?
8. Describe how a computer is used in placing orders. What else does a POS computer do?
9. How are items stacked properly on a tray or bus pan?
10. How should you present a check if the guest is to pay a cashier? If the guest were to pay the server?

Case Studies

Putting on a Banquet

The food and beverage manager of a large country club has a banquet in a room 120 feet by 100 feet. Using banquet tables that each seat eight guests, up to 1,000 guests can be served banquet style. A club president wants a banquet for approximately 800, plus or minus 5 percent.

The president wants the meal to be a first course, main meal with a salad, and dessert. He specifies the foods for each course. The food and beverage manager has a suggestion. After the main meal, omitting the dessert, the group would have its program, after which the guests would leave their tables and go to the large reception room next to the banquet room for dessert. The desserts would be displayed on four large tables, and the guests could select what and as much as they wish. They would eat standing up while talking to others. Coffee would be served at other tables nearby. He tells the president that this will allow him to lower the price of the banquet by a dollar per person. (He can afford to do this because it allows him to dismiss most of his service staff and keep only a skeleton crew to handle this dessert and coffee course.)

The president is doubtful. He is afraid of congestion and thinks that many of his club members will not like the arrangement. The food and beverage manager assures the president that he has not received complaints from this arrangement in the past. What other positive factors should the manager mention to convince the president that this is the way to go?

Handling a Personnel Problem

A large restaurant in a busy metropolitan area serves three meals a day. At lunch, it caters to shoppers, businesspeople, and high-income diners, mostly from surrounding hotels. It is a highly successful and thriving operation.

A large enrollment of downtown students at a business college near the restaurant has proven to be an excellent source of wait staff, both male and female. Careful selection is made to get neat, professional appearing, able female waitresses and male servers. Since the restaurant does a big business, the wait staff is large.

To fill an opening, a male waiter is hired. He's 23 years old with a dynamic personality. It is soon apparent that he is very lively and talkative. He immediately becomes of interest to a number of the female staff. He responds positively to this and becomes the envy of many of the male staff. In fact, quarrels and near fights break out. Some of the female waitresses also compete over him. What was a harmonious staff working together becomes a caldron of disharmony. It, of course, affects the service, and management begins to get more complaints than usual of poor service.

Management could correct the situation by letting the new hire go, but this may not necessarily rid them of the problem. Friction has developed that this would not correct. Besides, the new hire has developed into their best server. In the short time he has been there, he has developed a clientele, and patrons are asking for his tables. In fact, he's building business for the restaurant. Describe how you would handle this personnel problem.

Bar and Beverage Service

UTLINE

I. Introduction
II. Some Important Facts about Beverage Alcohol
   A. Alcohol Content and Proof
   B. Aging
III. Knowing Spirits
   A. Whiskey
   B. Liqueurs (Cordials)
   C. Gin
   D. Vodka
   E. Tequila
   F. Rum
   G. Brandy
   H. Grain-Neutral Spirits
IV. Knowing Wine
   A. Types of Wine
   B. Grape Varieties
   C. How Wines Are Named
   D. Tasting Wine
   E. Pairing Wine with Food
V. Knowing Beer
VI. Knowing Nonalcoholic Beverages
   A. Coffee
   B. Hot Chocolate
   C. Tea
VII. Hospitality Behind the Bar
VIII. Serving Spirits
   A. Serving Straight Drinks
   B. Serving Mixed Drinks and Cocktails
   C. Serving Cordials/Liqueurs
IX. Serving Wine
X. Serving Sparkling Wine
XI. Serving Beer
XII. Serving Nonalcoholic Drinks
XIII. Serving Alcohol Responsibly
XIV. Chapter Summary
XV. Chapter Review
XVI. Case Studies

Learning Objectives

After reading this module, you should be able to:

- Know enough to serve and recommend beverage alcohol and nonalcohol beverages to guests based on informed knowledge of beverages.
- Correctly follow legal and ethical procedures for serving alcoholic beverages.

INTRODUCTION

For thousands of years, people have concocted all kinds of alcoholic beverages. Many of these early brews were made from fermented milk, berries, and fruits. Ciders were made from every type of fruit known, and countless concoctions were created by mixing water with essences of herbs, seeds, and spices. Far older than written histories are cave pictographs showing humans drinking beverages in celebration. A several-thousand-year-old stone tablet was found on Mount Ararat, near where Noah's ark is thought to have landed, containing a recipe for making beer from grain. Another fermented beverage (tequila) was made from the agave cactus, and honey mixed with fermented liquid was used to make mead. Ciders were poured into open containers and allowed to stand until they "moved" by showing bubbles on the surface. The fermented juice of grapes was known to many early civilizations as far back as 8,000 to 10,000 years ago.

Coffee and tea were also consumed by early civilizations in southern Asia, where coffee was first discovered. Coffee, grown in the Middle East, Africa, and South America, has long been thought to have medicinal properties. However, Europeans in the eighteenth century found that when too much was consumed, it led to indigestion, depression, and states of irritability and anxiety. For these and other reasons, Pope Clement VIII declared coffee an infidel drink and forbade its use. (Later, another pope was tempted to taste it, did so, and found it much to his liking, and so he "baptized" it, thereby ridding it of its satanic properties.) In the United States, coffee was drunk as early as the 1660s. Today it is the most widely consumed beverage.

The Chinese were the first to drink tea made from the leaves and buds of a semitropical bush related to the camellia. It later became popular in Europe; tea has never been as popular in the United States as coffee is. Nevertheless, it is consumed in significant quantities.

Although Coca-Cola was first made in the 1880s, carbonated beverages were not mass produced until the 1920s and 1930s. Root beer, ginger beer, and ginger ale were the first, but with the introduction of cola, the popularity of carbonated beverages increased. Today cola drinks are popular all over the world. Americans drink a tremendous amount of carbonated beverages, and the amount per capita is increasing.

Specialty drinks with spirits, especially single-malt scotches and martinis, have made a comeback in this decade. Wine, especially Chardonnay, White Zinfandel, Cabernet Sauvignon, Chablis, and Pinot Grigio, has become very popular in the United States.

Premium and specialty beer sales grew in popularity throughout the 1990s and remain strong sellers today. Specialty beers from both large breweries and microbreweries have gained popularity for their distinct flavors. Many restaurants and bars are capitalizing on beer's increasing popularity by making more varieties available, promoting them through beer-of-the-month clubs, training employees in product

knowledge, introducing homemade beers through a brewpub concept, and suggesting beers with food. According to one recent study, the average beverage operation carries up to twenty bottled brands of beer and offers five brands of tap beer. Beer is by for the most popular alcoholic beverage. In 2003 it made up more than 50 percent of all alcohol sales.

FIGURE 9.1
The consumption of alcohol in social settings has a long tradition. Courtesy Corbis Digital Stock

Some Important Facts about Beverage Alcohol

The formation of alcohol is the result of an action brought about by yeast, which breaks down carbohydrates, freeing carbon dioxide, which adds to the brew a residue liquid called **ethyl alcohol.** It is a depressant and intoxicant. Alcohol is measured in people's systems as **blood alcohol content,** or BAC. Most people's liver can only break down about one ounce of alcohol per hour. Exceeding that amount in an hour can lead to dizziness, nausea, light-headedness, and more serious conditions. A BAC of as low as 0.05, or one-half drop of alcohol per thousand drops of blood, can begin to reduce inhibitions and reasoning, and slow down reactions. At 0.08 BAC, the legal limit in all fifty states for blood alcohol levels indicating intoxication a person might have slurred speech and blurred vision. At 0.15 there is a loss of muscle control, and alcohol can now act as a poison, causing vomiting and nausea. At 0.30 to 0.40, one passes out—a good thing, because from 0.40 to 0.50, the deep part of the brain is affected and one loses essential bodily functions, such as breathing and beating of the heart, leading to death.

Although some health risks are associated with overconsumption of alcohol, moderate consumption has been shown to have beneficial effects. It has been linked to lower risk of heart disease and lower blood pressure. Social drinking can lead to relaxation and conviviality. It can stimulate the appetite. Because it relaxes, it is often given to individuals just for that purpose. It expands the capillary system and makes the body feel warmer. It can help one sleep. There is evidence to show that those who drink alcohol moderately have fewer heart attacks than those who don't.

Alcohol Content and Proof

The federal government requires that the alcohol content of beverage alcohol be stated on the label of the container. In the United States it is stated in different ways. One is by weight, which we call proof. A spirit labeled 60 proof contains 30 percent alcohol by weight; one labeled 100 proof contains 50 percent alcohol by weight. The proof is double the alcohol content by weight. We do not indicate the alcohol content of wine and beer by weight, but by volume. This overstates the amount by weight of alcohol, because a liter of pure alcohol weighs 800 grams while a liter of water equals 1,000 grams. Thus, the ratio is 80:100 or 80 percent to 100 percent. In other words, a beer of 6 percent and a wine of 14 percent are respectively 4.8 percent and 11.2 percent alcohol ($0.06 \times 0.80 = 0.48$ and $0.14 \times 0.80 = 0.112$ or 11.2%). Other countries have different methods of labeling alcohol. The equivalent of an ounce of pure alcohol is two 12-ounce cans of beer, an 8-ounce glass of wine, 2½ ounces of 80-proof spirit, and 2 ounces of 100 proof spirit.

Aging

Some alcohol beverages, such as spirits and wine, are improved by age. Beer does not improve; in fact, it should be consumed as soon as possible after production. Wines age while in the cask and also after bottling. Spirits age only in the cask; after bottling there is little change. The date a spirit was made, indicated often on the label, can be a guide to how long is has been in the bottle. Some alcohol must be aged for certain periods of time to be called by a specific name; a *bottled-in-bond* spirit must be aged four years in a government warehouse. Servers dispensing beverage alcohol need to know what aging means in the various beverages they serve. Age can be an indication of quality, and guests might want to know the age to know the quality of the item ordered.

Knowing Spirits

Servers should know a great deal about the kinds of beverage alcohol they serve, so they can inform guests on what they are as well as how to serve them properly. The following sections will give you good basic information, but they are in no way complete. A server who wishes to be completely informed should take a special course in knowing and serving beverage alcohol.

Spirits are produced by the fermentation of grains, fruit, plants such as agave, sugar cane, and other products. After **fermentation** the brew is distilled at or above 160°F (71°C). The alcohol vaporizes, leaving the water in the brew. When this vapor is condensed it is a liquid made up of mostly alcohol. However, the alcohol carries over with it flavoring and other ingredients and so a number of alcohols take on the flavor of the products from which they were made. Others have flavors added to them after distillation. (See **Exhibit 9.1.**)

Spirits in the United States cannot be less than 80 proof; many are higher. However, cordials can be less.

A spirit labeled as **bonded** must be produced from a single distillation at 160 proof or less, be bottled at 100 proof, and be unblended, which means it is a straight liquor.

EXHIBIT 9.1 The Distillation Process

In the distillation process, the alcohol is vaporized at about 160°F (71°C), leaving the water in the brew.

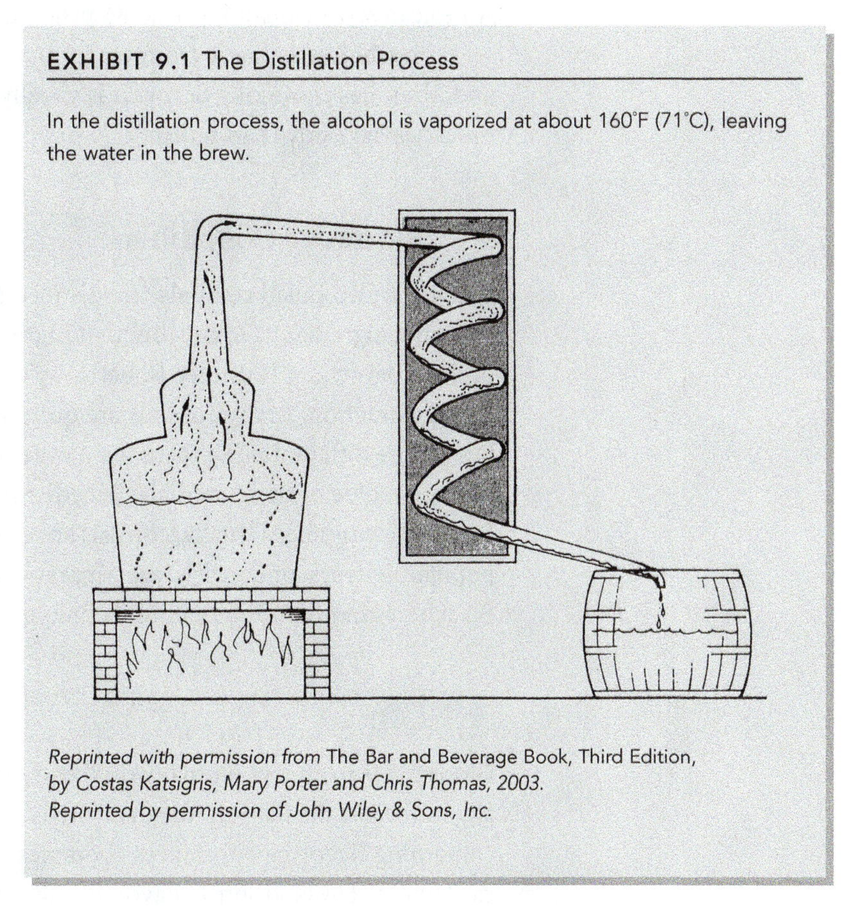

Reprinted with permission from The Bar and Beverage Book, Third Edition, *by Costas Katsigris, Mary Porter and Chris Thomas, 2003. Reprinted by permission of John Wiley & Sons, Inc.*

Whiskey

There are **straight** and **blended whiskeys**. Straight means that the spirit is from one distillation and has not been blended with grain neutral spirits. A blended whiskey is a straight whiskey blended with other batches of whiskey or grain neutral spirits. This ensures that all the whiskey from a particular brand has the same flavor and quality.

Bourbons must be made from at least 51 percent corn as the fermentable material and normally be aged four years in charred oak barrels at 125 proof. Kentucky bourbon must come from that state, but many bourbons come from other states. Rye whiskey is distilled from a mash containing not less than 51 percent rye as the fermentable material. It is like bourbon but is heavier in flavor. It must be aged at least two years in charred oak barrels. Corn whiskey must be made from a mash containing at least 80 percent corn. Canadian whiskeys are blended and can be made from rye, corn, or barley. They are lighter bodied than American whiskey, and with the trend toward lighter spirits this has made them heavy competitors of American whiskey. Canadian whiskey is made from liquors aged at least three years.

Scotch whiskey (or whisky, as it is spelled in Scotland) is made from malt or grain. Scotch's smoky flavor comes from drying sprouted barley over peat fires. American scotch must be 80 to 86 proof and aged at least four years.

Irish whiskey is made from barley but the sprouted grain is not dried over peat and, thus, has no smoky flavor. It is usually 86 proof and aged at least seven years; it has a heavier body than scotch.

Liqueurs (Cordials)

Liqueurs, also called **cordials,** are distilled spirits that have been treated with special flavoring ingredients. Many different ingredients are used to make a wide assortment of these beverages. Fruit, seeds, herbs, spices, honey, and other products are used for flavor extraction. Many cordials are quite sweet, which is an important distinguishing feature differentiating them from other spirits. Some cordials have been made for centuries. One of these cordials, Benedictine, is made from a secret, old recipe of the Benedictine monks. Usually, cordials are classified by their flavoring ingredient. Some popular liqueurs, or cordials, are Amaretto, Creme de Casis, Rock and Rye, Sambuca, Blackberry Liqueur, Peppermint Schnapps, Curacao, Maraschino, and Wisniak.

Gin

Gin is made from a spirit that would end up as a grain neutral spirit, except that usually in the last distillation, the distillate is allowed to come in contact with a gin head containing flavoring substances (botanicals) such as juniper berries, cassia bark, or cardamom. Holland gin is flavored only with juniper berries and is quite heavy in body and flavor. This makes it undesirable for making mixed drinks. London or English gin, also called dry gin, is lighter in flavor and body. It is a good mixing gin. Most American gins are similar to English gins. Few gins are aged; if they are, they take on a golden color and are called golden gin. Tanqueray, Booth's, Beefeaters, and Bombay are quite popular brands of gin.

Vodka

The federal government defines vodka as "neutral spirits, so distilled or so treated after distillation with charcoal and other materials, as to be without distinctive character, aroma, or color." Vodkas are typically between 80 and 100 proof, and they are not aged. Many countries around the world produce fine vodkas, such as Absolut (Sweden), Grey Goose (France) Stolichnaya (Russia) and Skyy (United States). Vodkas have enjoyed an increase in popularity due to the production of flavored vodkas using citrus, berries, and spices. Flavored vodkas such as Limonnaya, made from lemon peel, and Starka, made from brandy, vanilla, honey, and apple and pear tree scents, are uniquely flavored vodkas.

Tequila

Tequila is a distilled spirit made from the core of a nature blue agave plant. The starches in the agave plants core are turned to sugar through a roasting process. The juices are then extracted and placed in formatting tanks. Yeast is added to the tank, enabling the sugar to convert to alcohol. Tequila alcohol content is between 70 and 110 proof. A worm in a bottle of tequila is thought by many to be part of the tequila-making process. This, in fact, is a myth and is mostly a marketing ploy.

Rum

Fermented sugarcane or molasses are the sources of rum. **Light-bodied rums** (also called silver or dry rum) are quite smooth and have a light body and slightly sweet flavor. These rums are produced in Puerto Rico, the Virgin Islands, Cuba, Dominican Republic, and Haiti. Light rums usually have a white or silver label; they are usually aged one year. If aged, the label states **anejo** (old) or **muy anejo** (very old). The heavy-bodied rums, made in Jamaica, Barbados, Martinique, and Trinidad, are darker in color and richer in flavor. The label is usually tan or brownish. Bacardi, Ron Rico, and Rhum Saint James are all popular brands of rum.

Brandy

Any spirit distilled from a mash of fruit or fruit derivatives can qualify as brandy. However, only a distillate made from grapes can be called brandy; those from other fruits must be labeled by the name of the item used, such as Apricot Brandy. **Cognac** is a brandy that comes from the Cognac region in France; only brandy from this area may be called Cognac. **Armagnac** comes only from that French region. **Calvados,** an apple brandy, comes from Normandy.

Most brandies are aged, some for very long periods. Most brandies are 80–100 proof. Brandy labels often contain words, letters, or stars that indicate certain things; for example, **VS** or three stars means the brandies in the blend are less than 4½ years old, **Vieux** or **Grand Reserve** means most of the brandies in the blend are 20 to 40 years old and the youngest is more than 5½ years old.

Grain-Neutral Spirits

Many alcoholic beverages are made from grain-neutral spirits, a plain, unflavored liquid made of pure alcohol plus water in varying degrees of proof. If it is not made from grain, the neutral spirit cannot be called *grain.* A grain-neutral spirit is the alcohol used to make cordials, gin, vodka, blended whiskeys, and other products. If a label says "Blended Scotch" or "Blended Bourbon," then the straight—unblended—item has been diluted with grain-neutral spirits that have no flavor. The flavor is provided by the addition of the other distilled product.

Knowing Wine

Wine is made by fermenting fruit. In the process the yeast changes the sugars in the fruit to alcohol and carbon dioxide, while at the same time extracting the flavor of the fruit. Wines are not distilled. Aging can improve the flavor of wines. All wines are aged a year in the cask before they are bottled. After that, aging can continue. White wines age well, usually up to about six years, but some age longer. Red wines can age for many decades, but the change in flavor, color, and body, from the earliest to the latest wines, is usually dramatic. The alcohol content of wines can run from about 6 percent up to 20 percent or more. If above 14 percent, the wine usually has had a spirit added to it.

Types of Wine

Wine can be categorized into three types: table wines, sparkling wines, and fortified wines. Table wines are nonbubbling, or **still.** The CO_2 gas (carbon dioxide) that is a byproduct of fermentation is allowed to escape. Table wines usually have an alcohol range of 8 to 14 percent and are sold in restaurants and bars by the bottle, carafe, and glass.

Sparkling wines bubble because yeasts and sugar are added to still wine, causing a second fermentation in the bottle. These bubbling wines can be white, red, or rosé (blush), and usually contain 8 to 14 percent alcohol by volume plus CO_2. The most well-known type of sparkling wine is **Champagne.** For European wines, the name *Champagne* can only be used for sparkling wines made in the Champagne district of France. This district is in the northeastern corner of France and has a shorter growing season. Thus, the grapes are picked with higher acidity, which lends Champagne its distinct taste.

The following terms describe Champagne from the driest to sweetest:

- *Brut*—Driest
- *Extra dry*—Less dry
- *Sec*—More sweet
- *Demi-sec*—Sweetest

Brut and **extra-dry** are the Champagnes to serve throughout the meal. **Sec** and **demi-sec** are served with desserts and wedding cake.

Fortified wines are those with added alcohol, usually brandy. All wines with more than 14 percent alcohol have been fortified. Most fortified wines have an alcohol content of 17 to 22 percent. These wines come in two varieties: **apéritif** wines and **dessert wines.** Apéritif wines are flavored with herbs and spices, and are usually served before a meal. Vermouths from Italy and France are one type of apéritif; they can be sweet or dry. Dessert wines such as port, sherry, and Madeira, are typically sweet, rich, and heavy, so they are commonly served after dinner. The primary difference between port and sherry is that the brandy is added during fermentation with port for sweetness, and after fermentation with sherry.

Grape Varieties

Red wine is made from grapes with red, black, or purple skins. These grapes have been fermented with their skins on giving their color and **tannin**—the strong, slightly bitter flavor that makes the mouth pucker—to the wine. Red wines possess many hues—the exact shade is determined by the type of grape, the time the grape skins are left in the fermenting brew, and age of the produced wine. They tend to be heartier, heavier, and more flavorful than whites and rosés. In addition, red wines are almost always dry, or lacking in sweetness.

White wine can be made from red, green, or white grapes. The juice is extracted and separated from the skins. The result is then fermented into wine. White wines are lighter in color, flavor, and body than reds, and generally have a shorter life span. They can be very sweet or very dry, and range in color from pale yellow to gold; some may have a greenish tinge.

Blushes or **rosés** are pink and range in color from very pale to nearly red. They have a fruity flavor and are light and fresh; many have a slightly sweet taste. Thus, they resemble white wines more than reds. Blushes can be produced by fermenting grape skins for a shorter duration or blending red and white grapes. Some rosés are even called white, such as the white Zinfandel and white Pinot Noir wines from California. **Exhibit 9.2** shows various grapes.

How Wines Are Named

In the United States, wines are named in three ways:

1. By the predominant variety of grape used (varietal)
2. By broad general type (generic)
3. By brand name

Imported wines may also be named by their place of origin. For a **varietal** wine, the name of the grape is the name of the wine, and that grape will give the wine its

EXHIBIT 9.2 Grape Varieties

Different grapes produce many different types of wine.

Grape	Color	Body	Sweetness	Flavor Intensity	Region/Country of Origin
Barbera	red	medium to full	dry	medium to intense	California, Italy
Cabernet Sauvignon	red	medium to full	dry	medium to full	California, Bordeaux (France), Australia
Chardonnay	white	medium to full	dry	medium to full	California, Burgundy (France), Chablis (France), Champagne (France), Italy, Spain, Bulgaria, Australia, New Zealand
Chenin Blanc	white	medium	slightly sweet	medium	California, Loire Valley (France)
Gamay	red	light to medium	dry	delicate	Beaujolais (France)
Gewurztraminer	white	medium	dry	spicy, full	California, Alsace (France), Australia, New Zealand
Grenache	red	light to medium	dry	light to medium	California, Rhÿne Valley (France), Spain, Australia
Merlot	red	light to medium	dry	soft, delicate	California, Italy, Bordeaux (France)
Muller-Thurgau	white	soft to medium	sweet	mild to medium	Germany
Muscat	black or white		medium to sweet	medium to full	Italy, Alsace (France), Bulgaria
Nebbiolo	red	full	dry	intense	Italy
Pinot Blanc	white	light to medium	dry	light	California, Italy, Alsace (France), Champagne (France)
Pinot Noir	red	medium to full	dry	medium to full	California, Burgundy (France), Oregon, Champagne (France), Australia
Riesling	white	light to medium	slightly sweet	delicate	Germany, Alsace (France), Australia, California, Washington, Oregon
Sangiovese	red	medium to full	dry	medium to full	Italy
Sauvignon Blanc	white	medium	dry	medium	California, Bordeaux (France), Loire Valley (France), Chile
Sémillon	white	light to medium	dry	medium	California, Bordeaux (France)
Silvaner	white	light	dry	light	Alsace (France), Germany
Syrah, Shiraz	red	medium to full	dry	intense	Rhÿne Valley (France), Australia
Trebbiano	white	light to medium	dry	light to medium	Italy
Zinfandel	red	medium to full	dry	medium to intense	California

predominant flavor and aroma. Well-known examples include Cabernet Sauvignon, Chardonnay, Chenin Blanc, and Zinfandel. These terms were introduced to differentiate American wines from European wines. Varietals are very popular in this country. Legally, a varietal must include at least 75 percent of the dominant grape.

A **generic** wine is an American wine of a broad general type, such as burgundy or chablis. Their names are borrowed from European wines that come from well-known districts, which have a resemblance to the original. Federal law requires all American generics to include a place of origin on the label (such as California, Washington, Napa Valley). The best of the generics are pleasant, uncomplicated, affordable wines that operations often serve as house wines. As Americans have learned more about wine, they have come to recognize varietal names. Many wineries are using the terms *red table wine* or *white table wine* for these blends.

A **brand-name** wine may be anything from an inexpensive blend to a very fine wine with a prestigious pedigree. A brand name, also called a **proprietary name,** or in France, a *monopole,* is one belonging exclusively to a vineyard or a shipper who produces and/or bottles the wine and takes responsibility for its quality. A brand name alone does not tell you anything about the wine. The reputation of the producer should be your guide.

Tasting Wine

When tasting wine, a stemmed glass should be used, held by the foot or stem—never by the bowl, which would convey the heat of the hand to the wine. The same amount of wine should be poured each time (no more than half full) so that valid comparisons can be made.

There are generally considered to be five steps involved in the wine tasting experience.

1. It all begins with color. Holding the wine up to the light will help to show color, clarity, brilliance, and viscosity. Some terms that may describe a wine's appearance are: clear, bright, dull, hazy, or cloudy; thick/oily or thin/watery; pale or dark; weak or intense. Sparkling wines should have a steady stream of fine, abundant bubbles.
2. The next step is to swirl and smell the wine. Swirling helps the wine to contact the air, which develops its bouquet. Do not swirl sparkling wines, as this would release all the bubbles.
3. Smelling should be done with the nose inside the glass, taking deep breaths. Swirling and smelling can be done several times. If the scent is fruity or flowery, it is called an aroma—**aroma** reflects the scent of the young wine. As the wine ages, the aroma becomes the bouquet. The **bouquet** is the fragrance imparted by the winemaking and aging process. Descriptive words for the smell of wine include perfumed, woody, young, baked, complex,

closed, acrid, corky, moldy, skunky, minty, vegetal, and grassy. Many other familiar fragrances may be detected, including fruit, flower, and spice smells.

4. Actual tasting determines not only flavor, but a multitude of sensations including body, sweetness, acidity, and tannin. Tasting terms include syrupy, cloying, hot, biting, raw, harsh, mild, fat, nervous, round, firm, clumsy, flabby, velvety, soft, and generous. Also, the fruits, nuts, and spices and other flavors that were smelled can now be tasted. The wine can be either swallowed or spit out after tasting.
5. The impression left in the mouth is called the **finish.** Light, crisp wines will finish clean; great, complex wines will have a lingering finish. This is the evaluation period of a wine tasting.

Pairing Wine with Food

To match wines with food, you first need to taste the wines and foods in your operation so you can describe them enthusiastically and make recommendations. People's individual tastes vary, so nothing is cast in stone, but there are also some general rules you can follow.

FIGURE 9.2

The knowledge of how to pair particular wines with food selections will add to the dining experience. Courtesy PhotoDisc/ Getty Images

Appetizers—Dry to medium dry, light-bodied acidic white wines are usually good choices with appetizers, because they have refreshing quality that tends to stimulate the appetite. Sparkling wines are great combinations with many hors d'oeuvres.

Fish and seafood—Dry white wine (Chardonnay, Sauvignon Blanc) usually pairs best with seafood because the wine's crispness allows the food's subtler flavors to surface. Light-bodied red wines with little tannin (Gamay Beaujolais) are delicious with firm-fleshed fish such as swordfish.

Poultry and pork—Preparation, tastes, textures and appearances can vary. Carefully consider the sauce when pairing the right wine—a white, blush, or red may all be appropriate.

Veal—Leanness and delicate flavor is complemented by lighter red wines, well-aged red wines, or dry white wines.

Beef and lamb—Both are higher in fat and require wines with sufficient tannin to cut through full flavor. Complex red wines (Cabernet Sauvignon, Merlot) are good choices. Seasoned or marinated meats need a younger red (Pinot Noir).

Ham—Cured ham takes careful matching due to the saltiness or sweetness of meat. If the ham is glazed, a fruity rosé or blush (White Zinfandel) can be a nice match.

Exhibit 9.3 lists some more specific wine and food pairings.

EXHIBIT 9.3 Guidelines for Pairing Wine and Food

Certain foods are paired well with specific types of wine.

PASTA	
Fettuccine Alfredo	Frascati, Sauvignon Blanc
Lasagna	Chianti, Cabernet Sauvignon
Spaghetti primavera	Soave, Sauvignon Blanc
FISH/SEAFOOD	
Fish, grilled or broiled	Chardonnay, Sauvignon Blanc
Fresh shellfish	Sauvignon Blanc, Johannesburg Riesling
Fried catfish	White Zinfandel
POULTRY	
Barbecued chicken	Gamay Beaujolais
Sweet and sour chicken	White Zinfandel, Johannesburg Riesling
Chicken with light cream sauce	Chardonnay, Johannesburg Riesling
BEEF	
New York strip steak	Cabernet Sauvignon, Merlot
Beef Stroganoff	Merlot, Pinot Noir
Filet mignon with béarnaise sauce	Cabernet Sauvignon, Merlot
PORK/VEAL/LAMB	
Roast pork	Chardonnay, Gamay Beaujolais
Grilled pork chops	Gamay Beaujolais, Zinfandel
Veal parmigiana	Zinfandel, Chardonnay
LIGHT ENTREES	
Fruit salad	Johannesburg Riesling
Quiche Lorraine	Sauvignon Blanc, Chenin Blanc
Chicken Caesar salad	Sauvignon Blanc, Chardonnay

Knowing Beer

In today's bar and beverage operations, beer is the largest-selling beverage alcohol. In the United States, the first beer reference is from a Mayflower diary of 1622. The ship's captain had not planned to dock at Plymouth Rock, but he did when informed their beer was running out. They found they liked it there and stayed.

Beers are made from grains, yeast, water, and hops. Other grains such as rice and wheat are sometimes used. Malt made from sprouted barley is the grain most commonly used. Sprouting turns the barley's starch into a sugar called **maltose.** The sweetness of the malt is largely lost as the sugar changes to carbon dioxide and ethyl alcohol. However, in some beers the sweetness of the malt remains, such as in **porter** and **stout.**

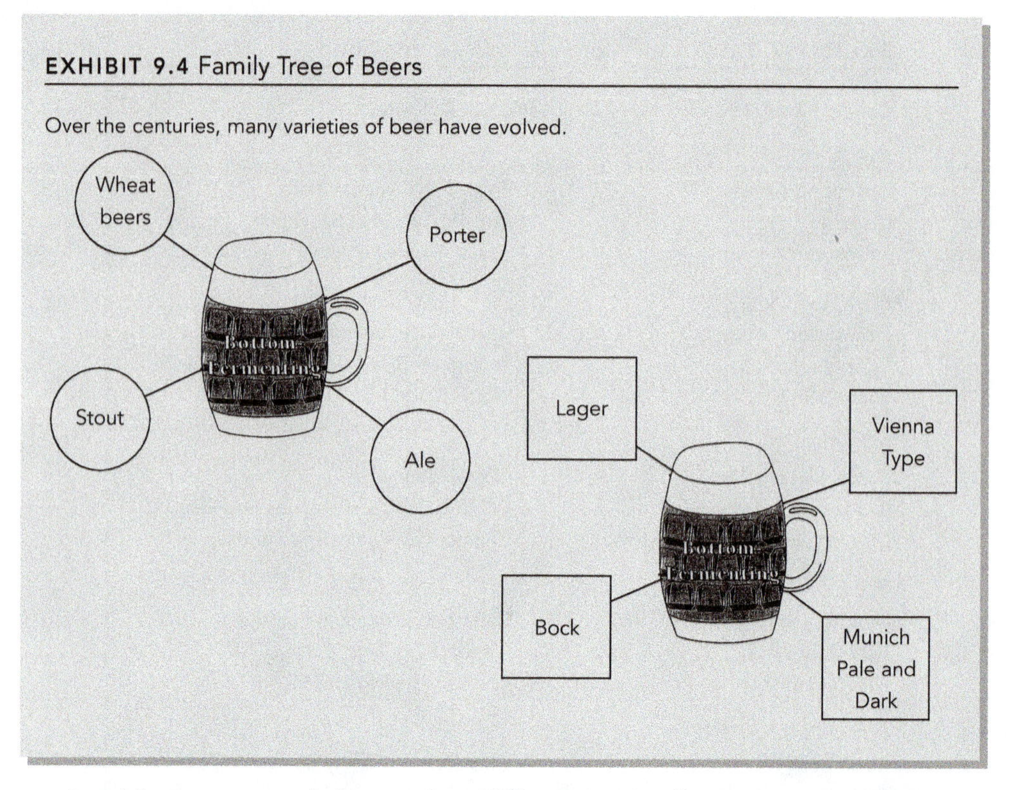

EXHIBIT 9.4 Family Tree of Beers

Over the centuries, many varieties of beer have evolved.

Special yeasts are needed to produce different kinds of beers. Beer is more than 90 percent water, and brewers often claim the excellence of their beer stems from the water used.

Hops give a pleasant bitter flavor to beer along with additional flavor. The most popular style of beer in the United States is **pilsner,** made by the **lager method,** which requires a long period of cold fermentation. Fermentation proceeds slowly from the bottom, so lagers are called **bottom-fermented** beers. Some other beers are **top-fermented,** which require only about three days to ferment. Ales, stouts, and porters are top-fermented.

Most lagers contain about 4 percent alcohol by volume. Ales may have a slightly higher alcohol content. Porter, stout, and **bock beer** have higher alcohol contents (about 6 percent) and are heavier in body, richer in flavor, darker, and sweeter than lagers. **Malt liquor** is highest of the beers in alcohol content.

Light beers are becoming increasingly popular due to their low-calorie content, which is about 100 calories per 12-ounce serving, as opposed to 140 to 175 for other beers.

Brewers are constantly striving for new types of beer that will attract the market. Steam beers, dry beers, and ice beers are recent successful entries into beer jargon. (See **Exhibits 9.4** and **9.5.**)

EXHIBIT 9.5 Guidelines for Pairing Beer and Food

Some beers are best consumed with certain types of food.

Food	Beer
Salads	Pilsner
Light appetizers	Lager, amber
Seafood	Lager
Spicy food, soup	Ale
Beef, game, other meats	Dark ale
Desserts	Stout, bock, double-bock, barleywines

Knowing Nonalcoholic Beverages

Coffee

Coffee has long played a major role in food service. The coffee tree (**Coffea Arabica**) is a tropical evergreen shrub native to Africa. The coffee shrub produces several types of beans, of which the **Robusta** and the **Arabica** are the best known. The word *coffee* seems to have originated from the Ethiopian region of **Kaffa,** although both the Turks and the Arabs claim to have named it first. The French botanical scholar and explorer **Jean de la Roque**, in his book *Voyage de l'Arabie Heureuse,* published in 1666, reports that the Turks made the beverage popular in the Middle Ages and called it **cahveh.** De la Roque was the first person to introduce coffee to Europe by bringing from his voyages coffee beans and brewing the first cup of coffee (Marseilles, France, 1644). The first coffee houses on records are found in the **Taktacalah** quarter of old Constantinople (now Istanbul).

Roasted coffees are sold as one type or blended. Light roasting produces a mild flavor, while long roasting creates a dark, stronger coffee. After roasting, the beans are ground. The fineness of the grind has much to do with the **extraction time,** or the length of time the water and the grinds are in contact. The extraction time for fine grinds is two minutes, after which used grinds should be removed to prevent the bitter remnants of the coffee from harming its flavor.

Coffee flavors are carried in bean oils. These oils are extremely fragile and oxidize very easily. To prevent off-flavor development within three days, ground coffee must be kept in a vacuum-sealed container or at freezing temperature. Whole-bean coffee keeps its flavor for about two weeks at room temperature. Some operations buy freshly roasted whole bean coffee and grind each batch independently to produce the finest quality coffee.

In the last two decades, decaffeinated coffee has become increasingly popular, so much so that in many foodservice operations patrons request it by a ratio of two to one. Decaffeinated coffee contains only 3 percent of the original caffeine. The caffeine is extracted by soaking the beans in hot water for several hours and by transferring them to a separate tank where they are treated with a solvent that absorbs the caffeine. **Trichlorethylene** and **methyline chloride** are the solvents most often used. However, the best method is by only water extraction, called the *European method.*

In the sequence of meal service, coffee is often served last and, thus, is charged with the responsibility of leaving the last impression to guests. In table service, unless requested earlier, coffee follows the entree, usually with or after the dessert. Before coffee service, the table should be cleared of all soiled items and unnecessary tableware. If the table has a tablecloth, it should be crumbed. In a coffee shop or family-style restaurant coffee is typically served in six-ounce cups. Coffee should be served hot—above 160°F (71.1°C)—to give the best flavor. Scrupulously clean equipment, good coffee and water, and proper brewing methods are required to make good coffee and, where servers are charged with its making, the principles for clean equipment and proper brewing should be observed. Coffee flavors are extracted best at about 205°F (96.1°C). Most operations today use a 10- or 12-cup coffee brewer, which permits service from the brewing pot. Decaffeinated coffee, espresso, cappuccino, caffé latté, café au lait, and other coffee drinks have become very popular and profitable.

Before serving coffee, use a serviette to wipe off the spout or mouth of the container. Alert guests that pouring is about to begin so they are aware of it and do not make a sudden movement that causes spillage and perhaps burning. The spout or container pouring edge should be from one to two inches above the rim of the cup and pouring should be done slowly and attentively. Many servers prefer to place the milk or cream container and sugar container on the table before pouring, but it can be done immediately afterward. They should be placed to the right of the cup. When serving at a table where there is more than one guest, it is proper to position them at the table center so it is easy for guests to help themselves.

FIGURE 9.3
Coffee should be poured from the right, with the handle at the three o'clock position. Courtesy PhotoDisc, Inc.

If the coffee is poured from a pot carried on a cocktail tray, the server, while pouring, must make certain that the tray is always level and balanced. When carrying a pot by hand, as when refilling a cup, it is recommended that an underliner be held on the other hand. Without it, there is always a risk of a drop or two falling on the table or onto a guest. All beverages are poured from the right and cleared from the right. The cup's handle should be placed at the three o'clock position with the teaspoon to the right of the saucer, pointed upward. While such positioning might seem trivial, to discerning guests it indicates consideration of the finer details of service. It usually takes the same amount of time and effort to place things properly as it does to place them incorrectly.

Hot Chocolate

Chocolate is derived from the bean of the semitropical **cacao tree.** This bean is roasted and broken into small pieces called **nibs.** The nibs are ground and the resulting product is bitter chocolate, a substance containing more than 50 percent cacao butter. Treating chocolate with **alkali** gives it a lighter appearance and smoother texture. It is called *Dutch* because it was first done in Holland.

Cocoa is produced after extracting some of the cacao butter (fat) from chocolate. A cocoa labeled *breakfast* contains 22 percent or more of cacao butter. Hot chocolate in most operations is a blend of cocoa, sugar, milk, and other ingredients.

To make hot chocolate from a mix, enough powder for one serving is placed into a cup, hot water is added, and the mixture is stirred. Whipped cream, shredded chocolate, and cinnamon may be served as toppings. The service of hot chocolate is patterned closely to that of coffee and tea.

Tea

Tea is an aromatic beverage obtained by infusion of tea leaves with boiling water. The word *tea* originated from the Chinese **T'e.** In the United States, tea has been widely consumed since the advent of the pilgrims. A historical event much related to tea consumption, the Boston Tea Party, was a factor leading up to the revolutionary war. To obtain a fine-quality tea, blends of different type of leaves are often used. The preferred teas are those made with leaves grown in higher altitudes.

In processing, all teas have their leaves first rolled to release juices and develop flavor. The next steps vary according to the type of tea to be made. **Green tea** is not fermented; in fact, it is often steamed or otherwise treated to prevent any change. It goes directly into drying after rolling. **Oolong tea** is lightly fermented and **black tea** is fully fermented and then both are dried. In the United States, an average of two pounds of tea per person is used each year. The best tea comes from the higher elevations.

Hot tea should be brewed in freshly boiled water above 185°F (85°C) for up to ten minutes. The custom of putting a tea bag into a cup and then pouring hot water over it and bringing the cup to the guest is not recommended. The tea should be put into a preheated stainless steel or ceramic pot and have boiling water poured over it. This pot should hold at least a pint so enough heat is produced to provide good extraction. After five minutes, the proper amount of extraction occurs, and guests should then remove the tea to prevent overextraction. The handle of the pot should point toward the guest's right. It is good service to also provide a pot of plain, freshly boiled water so the guest can brew the strength of the tea to that desired. Some guests prefer to add the bag to the tea while others wish the server to do this. Few desire cream for their hot tea but some like milk, while others just like a bit of lemon. The basic service procedures are the same as for coffee. In fine service, the milk or lemon is brought to the table on a plate lined with a doily. This is placed slightly above and to the right of the cup. Some operations offer exotic blends of brewed teas with added flavors such as spices, honey, herbs, fruits, or other flavorful ingredients.

Iced tea is usually made from orange pekoe black tea. The strength should be about double that of regular tea because of the dilution from ice. There is a danger of **clouding** in making tea of such strength when it chills. It also can cloud when tea is poured over ice. The reason is that chilling tends to precipitate the tannins in the tea, causing the clouding. For this reason, iced tea is usually newly brewed each day and is not refrigerated.

FIGURE 9.4
Many casual and fine-dining restaurants offer a selection of teas.

Iced tea is served in a tall glass with ice and a teaspoon. Sugar, brown sugar, regular sugar, or an artificial sweetener should be brought to the table prior to the service of the tea. Lemon is often served with the tea, preferably in wedges brought to the table on a small plate lined with a doily. Some operations find that using instant powdered iced tea pleases guests. Often the tea is in a portion package and the server only needs to add this to some cold water and ice to have iced tea.

Hospitality Behind the Bar

Excellent bar service relies on good organization, efficient control systems, and talented and knowledgeable staffing. A good bartender knows everything about what is for sale, how and when to suggest drinks, how to tell a joke or a good story, and how to serve responsibly. The bar can even become an information desk, especially in airports, train stations, and stadiums. The skills required of bartenders are many and varied.

Bartenders should come to work in a neat, clean work uniform. Before the shift, the bartender should check to see that everything is in order, neat, and clean. An inventory is usually taken of the storeroom before opening. Items are stored in their proper places. Supplies for the shift should be seen to be adequate. If an electronic liquor gun is used, the bartender should check where the liquor is stored to see that supplies are adequate and the system is connected correctly. Bar bottles should be checked to see that pouring caps are on, cordial corks are not sticky, and so forth. A sufficient supply of glassware should be on hand.

Offering food when serving alcoholic beverages is not only good hospitality but an important part of responsible service. In general, food slows the absorption of alcohol into the bloodstream. Furthermore, guests who are relaxing and having food may not drink as much or as quickly. This delay gives the liver more time to break down the alcohol in the person's blood. Food also helps protect the stomach from alcohol irritation. The foods that work best at slowing alcohol absorption are fatty and high-protein foods. These foods tend to take a while to digest, thereby keeping beverage alcohol in the stomach. Some examples include beef tacos, cheese, chicken wings, crackers and dip, fried foods, meatballs, and pizza.

FIGURE 9.5

A friendly bartender can be a bar's biggest asset. Courtesy Corbis Digital Stock

Advance preparation—mise en place—is one key to good bar service when things get rushed. Good organization and rhythmic preparation helps when a busy period hits the bar. In most times the bartender or bartenders are alone to handle business but it is good to have someone who can step in and help in times of stress.

Bartenders use house liquors or **well stock** to fill drink orders when no brand is specified. Well liquors are typically kept on lower shelves or in a well along the bar. Operations also offer **call stock,** brands that guests ask for especially rather than the regular well stock. Often these are premium brands and the charge for them is greater than for the well stock.

When mixing drinks in front of guests, the bartender places the glasses on the bar and mixes the drinks in a mixing container or pours from bottles. The bartender then puts down bar napkins, places each filled glass on a napkin, then gently slides each glass to the proper guest. Straws, stirrers, garnishes, or other accompaniments should be placed as the drink is put in place. Except for rind twists, garnishes should not be touched with the fingers, but put into place by using a decorative bar sword (a plastic toothpick in the shape of a sword) or toothpick.

Many guests do not like to see a bartender use a liquor gun to mix drinks because there is no flexibility in drink strength. Most operations allow bartenders to use free pouring, but this method makes it difficult to control portions and costs. Many bars portion drinks using **jiggers** that measure consistent amounts. When pouring with a jigger, place the glass on the bar in front of the guest and hold the jigger directly over it. Pour the liquor into the jigger, then dump it into the glass.

Standardized recipes that give the amount of alcoholic beverage, method of preparation, glassware to use, and garnish establish a strong foundation for customer satisfaction and profit. Customers should not be able to tell which bartender made their drink. Every drink must be the same. Method of service should also be standardized, and here also customers should not be able to identify which server served their drink.

Serving Spirits

Serving Straight Drinks

The term **straight** used with alcohol beverages means the spirit is served alone; nothing is added. If ice is served, the drink is termed **on the rocks.** Spirits, wines, and beers are usually served straight, although they also can be served blended with other ingredients. Straight spirits are typically served in an old-fashion glass (7-ounce) but also can be brought to the table in a smaller glass. Some operations like to bring either water, mineral water, or plain soda to the table with a straight spirit drink which, if the guest uses, makes it no longer a straight drink.

Serving Mixed Drinks and Cocktails

The term **mixology** refers to the art of mixing drinks. A mixed drink is any drink in which one beverage is mixed with one or more alcohol or nonalcohol ingredients. Mixed drinks include cocktails, highballs, tall drinks, frozen drinks, and coffee drinks. Many drinks have standard garnishes that are as much a part of the drink as the rest of the ingredients.

There are four basic mixing methods:

1. **Build**—Mix step by step in the serving glass, adding each ingredient one at a time. Built drinks include highballs, fruit-juice drinks, tall drinks, hot drinks, and drinks where one ingredient is floating on top of another.
2. **Stir**—Mix the ingredients by stirring them with ice in a mixing glass and then straining the drink into a chilled serving glass. The purpose of stirring is to mix and cool ingredients quickly with minimal dilution from the ice. Ingredients that blend together easily are stirred cocktails made of two or more spirits, or spirits plus wine.
3. **Shake**—Mix by shaking with a hand shaker or by mixing on a shake mixer (mechanical mixer). Shake drinks are made with ingredients that do not readily mix with spirits: sugar, cream, egg, and sometimes fruit juice.
4. **Blend**—Mix in an electric blender. You must blend any drink that contains solid food or ice. Strawberry daiquiris and frozen margaritas are blended. Although some bars blend drinks when they could easily shake or mix them, the blender is not nearly as fast as the mechanical mixer, and it does not make as good a drink as the hand shaker.

It is nearly impossible for one to know all of the various mixed drinks available today. Servers and bartenders should know the drinks usually requested by guests. If others are ordered, the bartender can look them up in a recipe book which should be kept at the bar. Bartenders should not guess or depend upon their memories, but should check. An improperly prepared drink not only displeases a guest but also results in a loss of ingredients when the drink is rejected and a replacement has to be made.

Serving Cordials/Liqueurs

Cordials, or liqueurs, are typically enjoyed before and after a meal as highly flavorful aperitifs and digestifs. Sweet cordials are a wonderful ending to a meal.

Liqueurs are usually served in small stemmed glasses that hold from 1 to 1 1/2 ounces or with ice in a larger bowled glass called a brandy snifter. In some fine dining restaurants, the cordial glass is served on a small plate with a doily and placed in front or slightly to the right of the guest. Often a well-appointed cordial cart is brought to the table by the captain or the server, thus enabling the guest to select among the various brands featured on the cart.

Serving Wine

Knowing how to serve wine can lead to increased sales for the operation and increased tips for the servers. Service need not be pretentious, but nothing can be more elegant during a meal than the opening and serving of a bottle of wine. Although wine is an excellent accompaniment to many foods, more and more guests are ordering it in place of other alcohol drinks. Today wine is served not only in food services but in practically every bar or other operation selling beverage alcohol.

Servers are also salespersons, and for them to do their best job in selling wines they should receive training on what wines are, their proper storage and service, and how to suggest wines that guests might like. This takes a considerable amount of knowledge, as well as serving skill. Wine and food have a long and historic association. Because of the wide variety of wines available, nearly every food can be paired with a wine that brings out its best qualities. Tradition held that red wines went with red meats, while white wines complemented poultry and fish, but this is no longer a set rule. The wine a person prefers is the proper wine.

Many find chilling a dry or medium-dry white wine increases its crispness, but overchilling can result in loss of bouquet and taste. White wines are best when served at 45°F to 55°F (7.2°C to 12.8°C) and reds at about 70°F (21.1°C). Wine is often served in chilled marble "coolers" placed on the table instead of a wine bucket. To chill wine quickly in an ice bucket, place a bottle in a, bucket filled half with ice and half with water. It takes about 15 minutes to bring a bottle of wine to about 50°F (10°C). If a wine bucket is filled with only ice, it is difficult to force the bottle down to the bottom; it also cools more slowly. When removing a bottle from the bucket, wipe it thoroughly. After guests have finished a beverage, remove the empty glasses promptly.

Some wines develop **sediment** (called **lees**) and should be decanted. Sediment can be left in the bottle by gently opening a bottle and then carefully pouring the wine slowly into a decanter from which the wine will be served. Some place a very short burning candle under the neck until they see the appearance of sediment.

In some operations a **sommelier** (wine steward) will give out the wine list, take the order, bring the glasses, and serve the wine. However, in most operations the server must do this. If guests have questions about what wine to order, the server should inquire about their tastes and preferences. When given the type of entree(s) ordered and specific price range requested, the server may suggest a number of appropriate wines to accompany the meal.

Follow the pivot system and use abbreviations when taking orders. After serving wine from a bottle, watch the table and be there before guests have to reach for the bottle and serve themselves. Be prompt in asking to bring another bottle after one is emptied or even before, if the bottle is nearly empty.

There are many glasses (practically all stemmed) designed for holding various kinds of wine; in some fine dining operations all of them will be used. This provides

FIGURE 9.6

Some fine-dining establishments have a sommelier to recommend and serve wine selections. Courtesy PhotoDisc/Getty Images

elegant service, but is costly and not necessarily appreciated by some guests. Many operations avoid the necessity for having a large inventory of expensive, stemmed glassware by using an all-purpose wine glass for all wines, including Champagne. However, Champagne and other sparkling wines are more often served in either a fluted glass holding about five ounces, or an aperitif or dessert glass holding three to four ounces. (See **Exhibit 9.6.**)

Servers should note that guests perceive the service of wine often as equal in importance to that of food; some even more so. In taking the wine order, the server should not make any adverse comment about the quality of the wine selected, the guest's pronunciation of the wine name, or anything else that might indicate a lack of knowledge on the part of the guest. The bottle should be brought to the table and presented unopened to the orderer, and held from the bottom with an opened hand while being held by the neck with the other hand. It is proper to hold a serviette or napkin behind the bottle about halfway around. This presentation is done so the orderer can check to see if what is being offered is what was ordered. Hold the label so it is at eye level, and give plenty of time so the guest can note the label's contents and make any comments desirable to the other guests. Handle bottles with care, being careful not to shake them, especially vintage reds, since this might disturb the lees and cloud the wine.

To open a bottle, rest the bottle on the side of the table between the person ordering the wine and the guest on the right, or turn to a nearby cart and open the bottle there. Cut the top foil away with the blade of the corkscrew, leaving the bottom foil around the neck. The cutting of the foil should be neat and quick. Wipe the top neck

EXHIBIT 9.6 Wine Glasses

A wide variety of wine glasses hold different types of wine.

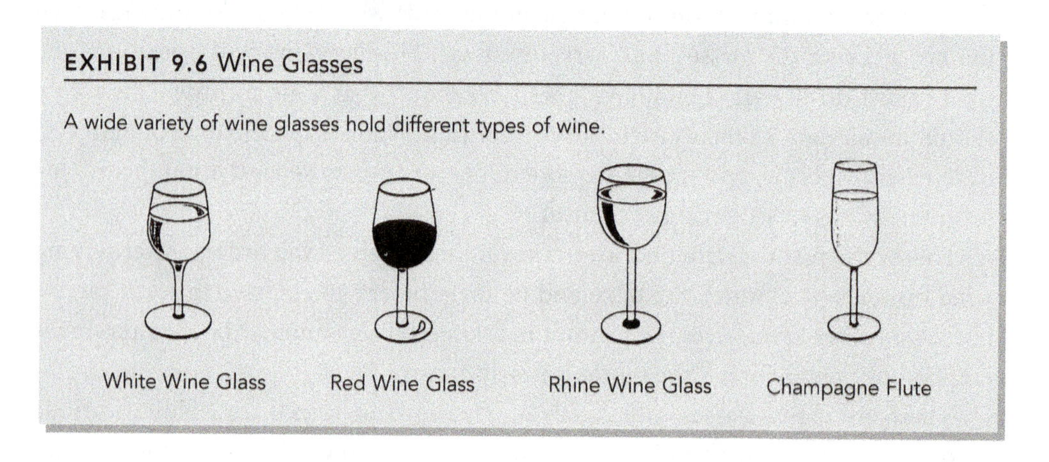

of the bottle and the top part of the cork with a napkin or serviette. Insert the opener into the center of the cork. Avoid placing the opener on the side of the cork; this makes it difficult to get a clean, smooth pull. Remove the cork.

The next step is to bring the bottle to the right of the person ordering and pour approximately one ounce into this person's glass, so it can be tasted and approved or disapproved. As the server ends pouring, the bottle should be given a slight, gentle twist with a short, quick movement of the wrist to prevent dripping. Next, remove the cork from the corkscrew and set it on a folded serviette or small plate to the right of the taster. Some may choose to smell the cork for the appropriate aromas and bouquets; it also should be felt, as a dry cork can indicate improper storage.

Serve the other guests in a clockwise manner, ending with the orderer. Some operations serve women before men. Fill glasses with still wine to about two-thirds full; sparkling wines tend to bubble up quickly when poured, leaving a bit of foam. A short pause in pouring allows this foam to subside and the glass may then be filled to the desired level. After pouring, place the bottle of white or sparkling wine into a wine bucket or insulated cooler (without ice) at table-side; rest opened red wines to the right of the orderer. Place older reds in a wine basket that holds the bottle on a slant so the lees are not disturbed.

FIGURE 9.7

Servers should be able to quickly and efficiently open a wine bottle. Courtesy PhotoDisc/Getty Images

Serving Sparkling Wine

When serving Champagne and other sparkling wines, it is very important to know the correct method of opening a bottle. A flying cork can sometimes be a dangerous missile. Make sure the bottle is chilled and has not been disturbed before opening it. The following steps outline the correct procedure for opening a bottle of champagne.

1. Cut the foil around the top of the bottle.
2. Place hand on top of the cork, never removing it until the cork is pulled out completely. Take off the wire.
3. Wrap a towel around the bottle for safety and spillage. Remove cork gently, slowly turning the bottle in one direction and the cork in another. Ease the cork out gently rather than jerking it out or letting it Shoot out from the bottle. Gentle removal does not disturb the wine and keeps the liquid from gushing out.
4. Pour the sparkling wine into the appropriate glassware.

Serving Beer

Beer may be served in a pitcher, in mugs, in a variety of glasses, or in a bottle accompanied by a glass. When serving, be sure to wipe the bottle before bringing it to the table. When a beer is poured at the table, place the glass or stein on the table to the right of the guest. If the server pours for the guest, he should pour carefully, not down the side of the glass, but in the center so about an inch of foam appears; it is proper to have the foam come to the rim. Some guests prefer to pour their own and will indicate so to the server. An alert server will remember this; if the guest is served again, the server will know that this guest prefers to pour his or her own. The amount of beer in different glasses varies, but 10- and 12-ounce glasses are common. Since there should be a collar of foam, the actual amount is not that of the full glass.

As the server finishes pouring from a bottle, the bottle should be given a slight twist with a quick movement of the wrist to stop drops from falling after pouring stops, Serving beer from a can is not recommended. However, the server can pour from a can into a glass, then place the can next to the poured glass for the customer.

Servers should serve draft beer within thirty seconds of its drawing. The beer looks unappealing if served flat; the foam should be between 1½ and 2 inches high. Draft beer ordinarily comes from a keg kept in a refrigerated area at about 40°F (4.4°C). It is carbonated as it comes out of the spout. Draft beer is very perishable because it has not been pasteurized to destroy harmful organisms. Canned or bottled draft beer is filtered through extremely fine filters to remove these organisms. This makes it unnecessary to pasteurize the product, which would spoil the draft taste.

Serving Nonalcoholic Drinks

Milk is usually served chilled and seldom with ice unless requested. The glass used is normally a tall highball, but goblets or plain 8-ounce glasses can also be used. Buttermilk, chocolate milk, and other complete milk drinks will be served in the same manner as milk. In the more casual dining operations, milk may be brought to the table in its paper carton resting on a small plate. Milk is often served to children in a plastic cup, with a lid and a straw.

Milkshakes, malted milks, and other milk and ice cream mixtures are served in tall, thick glasses topped often with whipped cream and a cherry. A straw and ice cream spoon should be served with them.

Juices are often served in a 6-ounce glass, but some large portions may be served in an 8- or 10-ounce glass. A highball glass may also be used. They should be served chilled but seldom are iced. In some operations servers pour their own juices; if so, attention must be paid to see the liquids are properly handled and refrigerated; they are highly perishable.

Although bottled waters have been around for a long time, they have increased in popularity in the last decade. They may be poured from the container and brought to the table or the opened bottle may be brought to the table and poured there by the server. Usually the glass used is the tumbler type; other shapes and sizes of glassware are used, depending upon the desires of the operation. When serving from a bottle at the table, it is proper to place the glass with ice on the table and pour by filling about a half of the glass. Sometimes a guest may request more and the server pours the glass to about three-fourths full. As with water, most guests dislike handling a glass filled nearly to the rim. In upscale operations the glass is placed on a small plate lined with a doily.

Carbonated beverages are often served to diners throughout the entire meal. Some guests will order sodas without caffeine and diet sodas. Some operations garnish certain drinks with a slice of lemon or other fruit. The service of carbonated beverages is the same as for bottled waters. If carbonated drinks are served from a soda gun they must be ordered from the bar. In some operations, servers operate a fountain themselves.

Serving Alcohol Responsibly

One has a greater chance of avoiding legal and criminal action and successfully handling problems related to alcohol beverages if there is a responsibly planned and continuously reviewed set of policies and procedures for every aspect of alcoholic beverage service. The following policies and procedures, in addition to those management thinks are important, should be in place:

1. Written policies and procedures include never serving a person alcohol to a point of intoxication, or serving an already intoxicated person, or serving minors.
2. Have good training program in alcohol service and in seeing those engaged in alcoholic beverage service are well trained in it, including disciplinary procedures for those who disobey them.
3. Eliminate certain marketing ideas, like 2-for-1s, and oversized drink promotions.
4. Check every young guest's ID to ensure minors are not served alcoholic beverages.
5. Offer food and nonalcoholic beverages.
6. Ensure that intoxicated guest do not drive away from the establishment. Call the police if one refuses assistance and does drive away.
7. Record all incidents in a log book.

The service of beverage alcohol is big business to the foodservice industry, but today their service to the public entails much more responsibility than it did in the past. This is because of numerous court decisions that have increased the liability of those engaged in serving alcohol to the public. Today those serving alcohol can be charged with responsibility for any mishap that might come to someone who becomes intoxicated from drinks served by them. The business in which the drinks are served can also be charged. If a person becomes intoxicated from drinks served in an operation and injures others because of the intoxication, both the server and the operation could be charged with what is called **third-party liability.** Heavy fines and even prison sentences are now possible.

Because of this greater responsibility, many states now require that those who serve alcohol take a course in alcohol awareness. These courses typically include training in the possible consequences of overserving, how to recognize signs of intoxication, how to tell guests they can no longer be served alcohol, and what to do in case someone arrives or becomes intoxicated. Servers must look for attitude changes, body language, speech, disruptive behavior, and any other deviation from normal behavior. The server, with support from a manager, must be trained to assess each situation and decide what action to take.

Alcoholic beverage service laws and their enforcement differ from area to area. The first rule of responsible alcohol service is to know and obey the laws that apply to the establishment. Currently, no federal agency sets standards for all beverage alcohol service, as the Food and Drug Administration (FDA) does for food safety and sanitation. Agencies such as police departments, departments of transportation, beverage alcohol commissions, and beverage alcohol and substance abuse associations can enforce local laws.

Most states use the BAC (blood alcohol concentration) test to determine intoxication (See **Exhibit 9.7.**) A BAC of 0.08 or more is considered intoxication in all states—this means that there are at least 0.08 drops of pure alcohol for every 1,000 drops of blood. A 150- to 180-pound person adds 0.02 to 0.03 BAC for every drink consumed—a drink is considered either two ounces of 80 proof spirit, 12 ounces of regular beer, or three ounces of 14 percent wine. About six ounces of 6 percent white wine would be about the same; this is why so many customers now ask for white wine—the alcohol content is lower. Four or five drinks brings about intoxication. A person below 150 pounds builds up a BAC of 0.04 for every drink, so three is a maximum. The normal body destroys about 0.04 BAC per hour. Thus, serving a 175-pound man five drinks in an hour should put the man in the **red zone.** (One system of monitoring intoxication puts a person with a low BAC in the **green zone,** one approaching intoxication in the **yellow zone,** while one who is in danger of being or is intoxicated is said to be in the red zone.) The reason the man having five drinks in an hour is in the red zone is that if each drink contributes 0.025 BAC, he has had enough alcohol to build a BAC of 0.125 (1 x 0.025), but his body has probably destroyed 0.04 BAC and so he should have a BAC of 0.085 (0.125 – 0.04). This is close enough to 0.10 for the server to take cautionary measures.

EXHIBIT 9.7 BAC Chart

BAC, or blood alcohol level, indicates how intoxicated a person is.

Body Weight	Number of Drinks* during a two-hour period									
100 lbs	1	2	3	4	5	6	7	8	9	10
120 lbs	1	2	3	4	5	6	7	8	9	10
140 lbs	1	2	3	4	5	6	7	8	9	10
160 lbs	1	2	3	4	5	6	7	8	9	10
180 lbs	1	2	3	4	5	6	7	8	9	10
200 lbs	1	2	3	4	5	6	7	8	9	10
220 lbs	1	2	3	4	5	6	7	8	9	10
240 lbs	1	2	3	4	5	6	7	8	9	10
	Be Careful Driving BAC to 0.05%			Driving May Be Impaired 0.05–0.09%**				Do Not Drive 0.10% and up		

BAC—Blood Alcohol Concentration

* *One drink is 1 1/4 oz of 80-proof liquor, 12 oz. of beer, or 3 oz. of 14% wine.*
** *In all states, legal intoxication is 0.08%.*

This chart provides averages only. Individuals may vary and factors, such as food in the stomach, medication, and fatigue, can affect your tolerance.

Courtesy of the National Restaurant Association.
Source: Distilled Spirits Council of the United States, Incorporated.

Carefully check IDs. Many establishments check IDs of anyone who looks younger than 30. Clubs or bars patronized by young people often check everyone's ID. In either case, be sure that IDs are valid and depict the person presenting the ID. Minors frequently obtain alcoholic beverages by presenting someone else's ID or a fake one. Set up a complete ID-checking station at each entrance and staff it with employees trained in this task.

To verify customer identification if customer requests an alcoholic beverage, take these steps:

1. Ask for identification from customer.
2. Examine identification.
   - Verify photo is that of customer.
   - Verify date of birth.

- For identifications in question, compare to ID guidebook:

  Problem Possibility:
  Customer does not look like photograph.

  Problem Response:
  Obtain other form of identification.

  Problem Possibility:
  Customer does not have any identification.

  Problem Response:
  Politely refuse to serve alcohol.

  Problem Possibility:
  Customer does not have a driver's license.

  Problem Response:
  Examine other acceptable form of identification.

  Problem Possibility:
  A dispute of validity of the identification of a customer arises.

  Problem Response:
  Contact management and brief manager or head waiter on situation.

3. Thank customer and return identification.
4. Check everyone in the party who is questionable.

State liquor codes govern liquor licenses. One should establish a good relationship with law enforcement and regulatory agents in your area by adhering to your state liquor code and practicing good business ethics. Disobeying liquor code regulations is a criminal violation that can result in fines, imprisonment, and suspension or loss of your liquor license.

Each area's liquor code covers a variety of regulations, but the key issue is who not to server. In most areas it is illegal to server alcoholic beverages to the following people:

- Minors under 21 (All guests must be able to show proper picture identification establishing their age.)
- People who are already intoxicated

There is no penalty for refusing to serve someone you merely suspect is a minor or an intoxicated person. You have the right to protect your guests, yourself, and your establishment's liquor license.

CHAPTER SUMMARY

Beverage service requires specialized knowledge. Beverage alcohol has enjoyed a long and distinguished history, and has played an important role in human religious, social, and philosophical development. Today, restaurants and hotels may find a third to half of their net profit comes from beverage alcohol service.

Alcohol is produced by the fermentation of a carbohydrate, which is changed by the process into alcohol and carbon dioxide. The amount of alcohol is stated in terms of proof, which is twice the percentage of its pure alcohol. The alcohol content of wines and beers is stated in percent of volume. Because alcohol is lighter than water, it takes more to make the same weight as water. Thus, a spirit of 12 proof, which is twice its real alcoholic content or it is actually 6 percent by amount or weight, has less alcohol than a wine or beer of 6 percent in volume. Put this way, if it is a jigger of alcohol versus a jigger of water, the jigger of water is greater than the jigger of alcohol. It takes just a bit more alcohol to make the same weight. Thus, a wine or beer of 6 percent by volume has less alcohol than a spirit of 6 percent by weight.

Spirits include whiskeys, gins, brandies, vodkas, rums, and grain-neutral spirits. Grain-neutral spirits are used as the base of cordials, vodkas, gins, and for blending into whiskies.

Wines are categorized into three types: table wines, sparkling wines, and fortified wines. Tables wines are still. Sparkling wines, including Champagne, are bubbling with carbon dioxide. Fortified wines have added alcohol, usually brandy. Fortified wines include apéritifs, vermouths, port, sherry, and Madeira.

Red grape varieties include Barbera, Cabernet Sauvignon, Gamay, Grenache, Merlot, Nebbiolo, Pinot Noir, Sangiovese, Syrah (or Shiraz), and Zinfandel. White grape varieties include Chardonnay, Chenin Blanc, Gewurtztraminer, Muller-Thurgau, Muscat, Pinot Blanc, Riesling, Sauvignon Blanc, Semillon, Silvaner, and Trebbiano. Wines may be named for their predominant grape, generic type, or brand name.

The five steps of tasting wine are looking, swirling, smelling, tasting, and finish. Although there are general guidelines for pairing particular wines with foods, the most important consideration in a restaurant is the guest's preference.

Beer is made from grains, yeast, hops, water, and and a fermentable product called an adjunct. Types of beer include lagers, pilsners, ales, stouts, porters, malt liquor, bock beer, light beers, steam beer, dry beer, and ice beer.

Coffee, tea, and hot chocolate service are very much the same. Servers must use precaution when serving these hot beverages. The two best known coffee bean varieties are Robusta and Arabica. Well-brewed coffee should be served at about 160°F (71.1°C) and held no longer than one hour. Popular coffee drinks include espresso, cappuccino, caffé latte, and café au lait. Teas come in a variety of colors and flavors, and are best served in a pot. Servers should know the appropriate garnish for the wide variety of hot beverages available.

Bar service requires many skills of bartenders besides the ability to mix and serve drinks. Speed, precision, service skills, and attention to good sanitation are equally important. Basic knowledge behind the bar includes the ability to mix a wide variety of drinks, as well as knowing the difference between well liquors and call brands.

Spirits may be served straight up, on the rocks, mixed with one mixer, or as a cocktail. Drinks may be built, stirred, shaken, or blended. Brandy is traditionally served in a snifter.

Review the steps of serving wine carefully. The guest dining experience depends greatly on the impression left by the server's grace and speed in presenting, opening, and pouring wine. The server must be able to suggest wines and answer questions about any wine on the list. Extra care is required when serving sparkling wine so that the cork does not fly out of the bottle.

Beer must be served with a healthy head, which is accomplished only by pouring from the bottle or draft-beer system at the correct angle.

The legal climate surrounding alcohol service has changed drastically in the last decade as third-party liability has become more common. All servers of alcohol must be trained in responsible service, which includes knowing how alcohol affects the body in terms of blood alcohol content (BAC), how to monitor and control guests' drinking, and what to do when a guest becomes intoxicated and tries to drive away. Remember that a full training program in this area is required in most states, and thorough training helps protect an operation in case of a third-party lawsuit.

Key Terms

alkal
anejo/muy anejo
apéritif
Arabica
armagnac
aroma
black tea
blended
blood alcohol content (BAC)
blushes
bock beers
bonded
bottom fermented
bouquet
brand-name
brut
cacoa tree
cahuch
call stock
calvados
Champagne
clouding
Coffee Arabica
cognac
Cordials
demi-sec
dessert wines
ethyl alcohol
extraction time
fermentation
finish
fortified
generic
Grand Reserve
green tea
hops
Jean de la Roque
jiggers
Kaffa
lager method
light-bodied rums
liqueurs
malt liquor
maltose
methyline chloride
mixology
nibs
on the rocks
oolong tea
pilsner
porter
proof
proprietary name
red, yellow, and green zones
Robusta
rosé
sediment
sparkling wines
spirits
still
stout
sommelier
straight
Taktacalah
tannin
T'e
tea
third-party liability
top fermented
trichlorethylene
varietal
Vieux
VS
well stock

CHAPTER REVIEW

1. How does aging affect spirits, wine, and beer?
2. What is BAC? At what level of BAC do most states say one is intoxicated? About how much alcohol does the ordinary person break down per hour?
3. If a spirit is 80 proof, what is its alcohol content?
4. How is bourbon different from other types of whiskey?
5. What is used to make rum?
6. How high should wine glasses be filled?
7. List the qualifications you would want in hiring a bartender.
8. What is well stock? What is a call brand?
9. A party of six comes in for service, two of whom appear to be minors. What should the server do?
10. Describe how to pull the cork out of a bottle of wine.
11. Taste a Scotch, an Irish whiskey, a bourbon, and a rye. Write one word to describe the flavor of each.
    (1) Scotch: ____________________
    (2) Irish whiskey: ____________________
    (3) Bourbon: ____________________
    (4) Rye: ____________________
12. Taste each of the following spirits, and write one word to describe the flavor of each.
    (1) Sambuca: ____________________
    (2) Curacao: ____________________
    (3) Gin: ____________________
    (4) Vodka: ____________________
    (5) Rum: ____________________
    (6) Cognac: ____________________
    (7) Tequila: ____________________

13. Set up four plates of food: one salty, one with red meat, one with fish, and one with a cream sauce. Taste each food along with a sip of each of the following wines. Describe the flavor of each combination.

(1) Sparkling wine: ____________________

Salt: ____________________

Meat: ____________________

Fish: ____________________

Cream: ____________________

(2) Cabernet Sauvignon: ____________________

Salt: ____________________

Meat: ____________________

Fish: ____________________

Cream: ____________________

(3) Chardonnay: ____________________

Salt: ____________________

Meat: ____________________

Fish: ____________________

Cream: ____________________

(4) Merlot: ____________________

Salt: ____________________

Meat: ____________________

Fish: ____________________

Cream: ____________________

(5) Pinot Noir: ____________________

Salt: ____________________

Meat: ____________________

Fish: ____________________

Cream: ____________________

14. Set up four plates of food: one salty, one with red meat, one with fish, and one with a cream sauce. Taste each food along with a sip of each of the following beers. Describe the flavor of each combination.

(1) Stout: ______

Salt: ______

Meat: ______

Fish: ______

Cream: ______

(2) Porter: ______

Salt: ______

Meat: ______

Fish: ______

Cream: ______

(3) Weiss: ______

Salt: ______

Meat: ______

Fish: ______

Cream: ______

(4) Pilsner: ______

Salt: ______

Meat: ______

Fish: ______

Cream: ______

(5) Pale ale: ______

Salt: ______

Meat: ______

Fish: ______

Cream: ______

15. Invent a new coffee drink that contains at least three ingredients, including a garnish. Write the recipe here.

____

____

____

____

____

____

____

____

____

____

____

The Ideal Server

You are to present your ideas to the class on what you feel is the ideal server. Write a list of the things you feel are important. Be sure to include ideas on serving people with various handicaps, handling alcohol problems, and mollifying difficult guests who try to raise an argument.

The Award Problem

The manager of a white tablecloth restaurant decides to give an award to the server who sells the most wine in a two-week period. She decides to give different points for various priced wines with the server with the highest score winning $500. She publishes the following price and point schedule for the contest:

Wine Price	Points
Under $10.00	10
$10.01–$15,00	15
15.01–20.00	20
20,01–25.00	25
25.01–30.00	30
30.01–35.00	35
35.01–40.00	40
40.01–45.00	45
45.01–50.00 and over	50

The contest was a spirited one and enjoyed by all the servers. There was much camaraderie and general good will developed by the servers. Mary Jones was declared the winner with 134 bottles of wine sold and 2,808 points. Joe Tobias was a close runner-up, with 122 bottles sold for a total of 2,788 points. The rest of the server staff did fairly well, but these two were the leaders. When the manager found the contest boosted sales by 12.1 percent, she considered the contest very worthwhile.

A few days later, Joe Tobias made his own check of his total bottle sales and points. He found he sold 126 bottles of wine with total points of 2,815. He called management's attention to what he believed was a mistake in the calculations. The manager checked the figures herself and found that the person who made the original calculations had indeed made a mistake. Tobias was the winner.

What should management do? Nothing? What would be the results if this were done? Should she ask Mary Jones for the $500 back and give it to Joe?

Should she instead call it a tie and also give Joe $500? Is this a good or bad way to solve the problem? Explain your reasoning.

The manager realizes she has a morale problem here. The wrong decision could destroy staff morale and the chances of ever running a successful contest again.

Critique this contest on the basis of using this point system for deciding the winner. Was there an easier and simpler way? Was it a good idea to give only a first prize? What would you have done? Why?

Set up your own rules for running such a contest.

MANAGEMENT'S ROLE IN SERVICE

OUTLINE

I. Introduction
II. Establishing Service Standards
III. Management Functions
 A. Planning
 B. Organizing
 C. Staffing
 D. Leading
 E. Controlling
IV. Motivating Servers
 A. Hiring Motivated Servers
 B. Creating a Motivating Environment
 C. Empowering Employees
 D. Rewards as Motivators
 E. Evaluations as Motivators
V. Scheduling Servers
VI. Dining Room Arrangement
VII. Kitchen Arrangement
VIII. Training Servers
 A. The Effective Trainer
 B. The Training Session
 C. The Training Space, Tools, and Equipment
IX. Reservations
X. Chapter Summary
XI. Chapter Review
XII. Case Studies

LEARNING OBJECTIVES

After reading this chapter, you should be able to:

- Describe management functions necessary to a successful operation.
- Explain how motivating and training servers helps an operation deliver excellent customer service.

Introduction

Good service comes not only from servers, but also from managers. Managers bear a great responsibility for establishing service standards, motivating servers, scheduling servers, training servers, and providing them with the equipment, tools, and environment they need to do their job well. The establishment of good service is a partnership task between managers and servers working as a team to deliver it.

To have good service, management must set high service standards, communicate them to employees, see that they are met or exceeded, and support employees in their efforts through training and recognition. Unless this is done, servers are apt to establish their own standards and practice them. This may not give the best in service.

Establishing Service Standards

Standards can be defined as specific rules, principles, or measures established to guide employees in performing their duties consistently. With standards, management can measure and evaluate employee performance and operation performance toward pleasing guests.

Service standards include service policies and service mechanics models, such as how to pour water, how to deliver food to the table, how to set up a table, or how to prepare a service station. Management establishes service standards based on the type of operation and the quality of service management wants to achieve. In a fine-dining environment, the service standards for wine service are usually more extended and elaborate than they are in an informal, family-style operation (see Chapter 9).

The first step in managing and delivering quality service is to set up and communicate service standards. Setting service standards involves seven distinct steps by managers:

1. Set standards and describe them in detail.
2. Establish policies and procedures for accomplishing these standards. A *policy* is a plan or course of action to meet a standard. A *procedure* is the manner in which that plan is implemented.
3. Provide the necessary space, equipment, and environment to achieve standards.
4. Provide adequate training and guidance to servers and ensure that standards are met.
5. Review with employees periodically, so they know and understand how standards performance will be measured.
6. Train employees to perform specific tasks to meet standards. Follow up with checklists, sidework, tasks lists, and job descriptions.

7. Encourage and seek out employee feedback so management can be aware of problems that need correction.

Experienced managers are unanimous in reporting that the establishment of standards is a relatively simple task. The most difficult task is to train employees in those standards and then see that they follow them consistently day after day. Often, management establishes the standards and then delegates to secondary management the task of seeing that they are followed.

Standards vary, from the elegant service required in a fine-dining establishment to that appropriate for casual dining facilities. This book discusses typical full-service standards, but every operation must interpret these to suit its guests. In other words, what is correct and what is not correct will depend on the operation and situation. Specific standards have to be built for each operation, and it is management's responsibility to establish them.

Management Functions

Good service occurs when operations are well managed. A poorly run organization is unsettling to employees and leads to poor productivity and work performance. Seeing that an operation is properly run is a management responsibility and revolves around managers performing **five management functions:** planning, organizing, staffing, leading, and controlling.

Planning

The planning task starts with establishing a mission and goals for the organization and its people. This should be done with input from all employees, including servers. After all, who knows customers better than the people who serve them every day? All employees must have a stake in organizational goals or they won't be realized. Once goals are set, managers must see that realistic plans are developed, followed, and revised so these goals are met.

Organizing

Organizing applies to all of an organization's processes and resources, including its people. An organizational chart shows an operation's positions and their relationships to each other, including who reports to whom. (See **Exhibit 10.1.**) When responsibility and authority in an organization flow from the top down, it is known as a **line organization.** In practice, a hostess may be responsible for servers and to an assistant manager, who reports to a manager. A number of head waiters may be responsible to the hostess, and each head waiter may have a group of servers responsible

EXHIBIT 10.1 A Sample Organization Chart

An organizational chart displays an operation's positions and their relationships to each other.

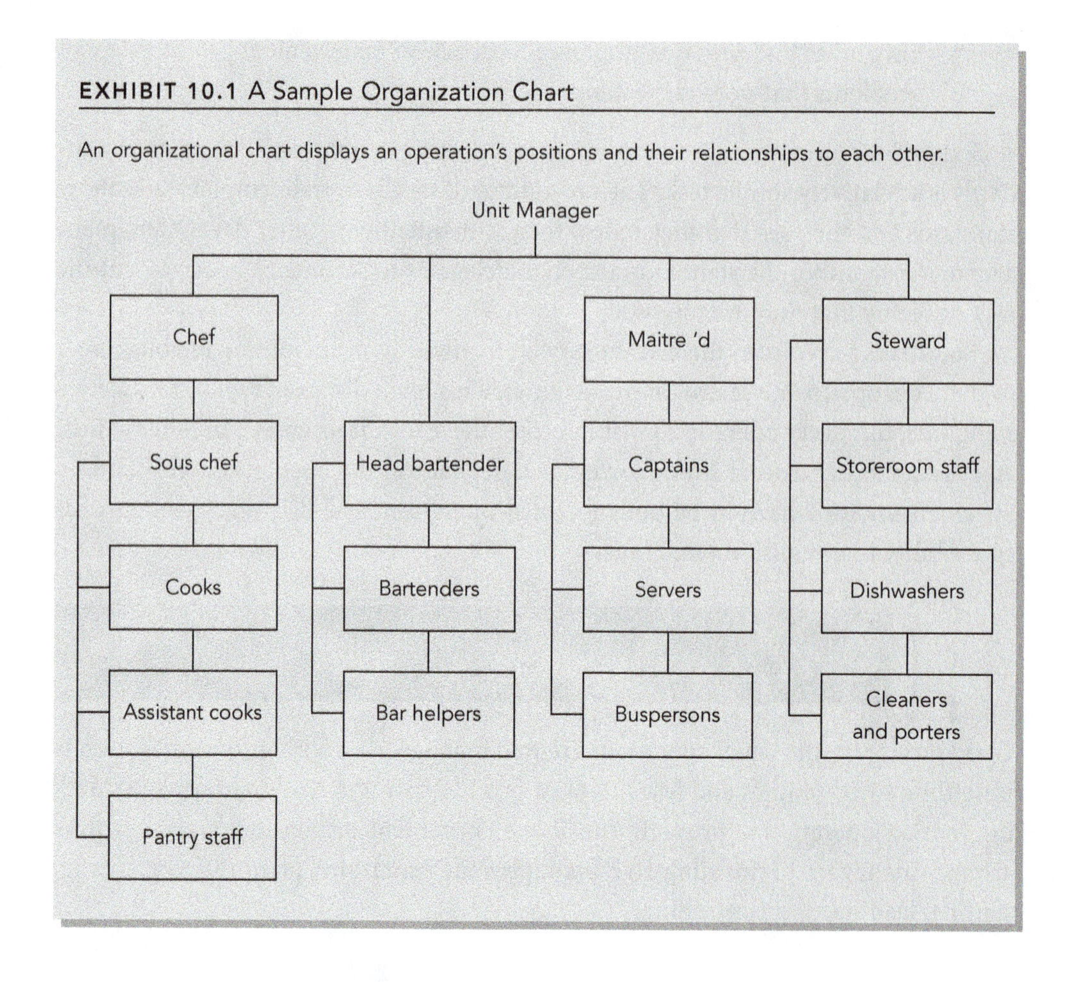

to him or her. Large hotels tend to use a complex line organization, from food and beverage managers, to maitre d', to dining room captains, to chefs de rang (food servers), to commis (assistant servers), to buspersons. An important factor in line organization is that people should have to report to only one immediate superior. Reporting to more than one person and circumventing proper communication channels can cause needless frustration and miscommunication. For example, *unity of command,* as defined by management theory, ensures that all employees throughout the operation follow the same policies and regulations.

If a number of employees have a problem with their supervisor, they should be encouraged to talk to another person with authority to effect results.

Many organizations are replacing line organization with an approach called *team effort.* One version of this is the **reverse pyramid** in which the manager is at the bottom of the pyramid, supervisors are in the middle, and front-line employees are on the top. In this model, managers are seen as serving front-line employees so they can better serve customers (see **Exhibit 10.2**).

An important component of organization is **delegation.** Responsibilities and tasks can be delegated to employees, but the delegator does not escape responsibility for seeing that the job is done properly. An important part of delegating responsibility is also to give an employee authority to accomplish the work. Without authority, employees' hands are tied.

Proper delegation can work to the advantage of management:

- It can help the manager perform better on the job.
- It can be instrumental in establishing a definite sense of pride in the employee charged with the new responsibility.
- It can save managers time by freeing them from some tasks that will allow them to give that time to other important tasks.

Proper delegation requires the following:

- The delegatee must know how to do the delegated tasks.
- The person being delegated to do a task must be willing to do it.
- The person delegating assignments must provide the resources and all the assistance needed to do the job. The delegatee must be given the authority to do the tasks. If this is not given, usually the delegation tasks are not done or are improperly done.

EXHIBIT 10.2 A Sample Reverse Pyramid Organization Chart

In the reverse pyramid, managers are at the bottom, with front-line employees and customers at the top.

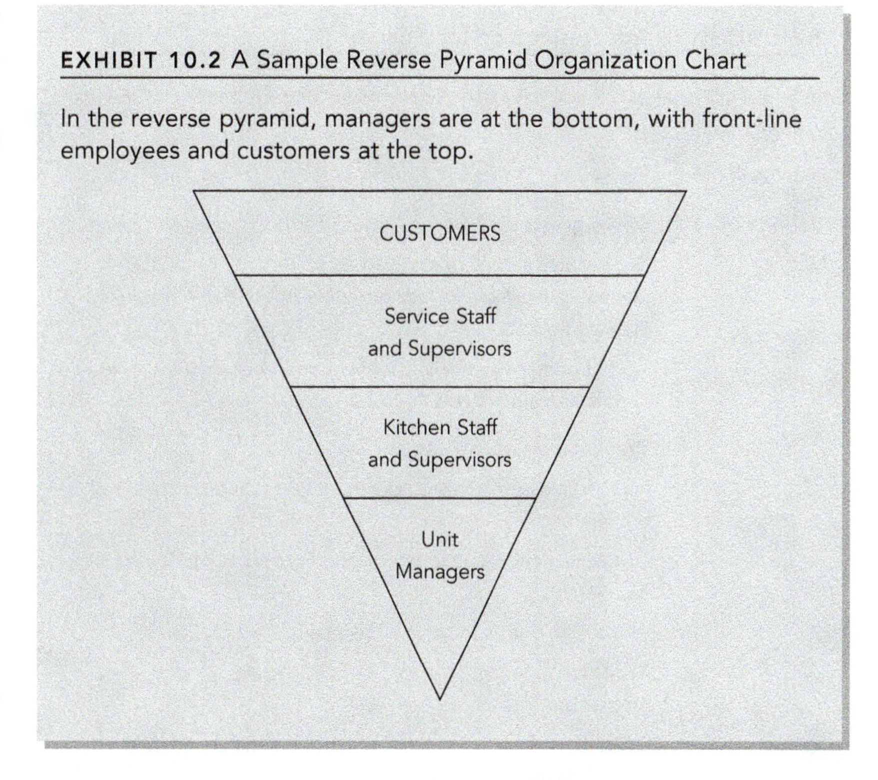

Staffing

Hiring qualified employees is the first step in creating an excellent service staff. Hiring an employee that knows what is expected and will meet operation standards is more important than hiring an employee that simply acknowledges the specific duties of the job. Just because an employee has done the job before does not mean they will meet the hiring operation's standards. The person doing the hiring must ascertain whether those seeking positions have the necessary abilities and attitudes for the job, and reject those that do not have the necessary attributes.

One builds a staff by first determining what tasks must be performed to achieve the desired goal and then allocating positions so that these tasks are performed. After this, job specifications are written up, indicating what the person in each position does. **Exhibit 10.3** is an example of a job description written up for a server. The next step is to hire people who can adequately fill these positions.

A staff consists of a group of individuals who are working together toward a common goal. A number of specialized positions are needed to perform the various tasks.

EXHIBIT 10.3 Job Description for a Server

Job descriptions indicate the necessary requirements for each position in the organization.

Job Title	Server
Job Summary	Serves guests in all ways to ensure maximum sales and profits
	Coordinates service of meals to customers to ensure 100 percent satisfaction, including troubleshooting and resolution of all complaints
	Acts as primary bridge of communication between the operation's management and its customers
Job Objectives	Greets guests warmly and sincerely
	Informs guests of specials, signature items, and specialties of the operation
	Suggests menu items to guests and answers questions about the menu
	Works with guests to solve all service-related problems
	Keeps manager informed of potential and actual service problems
	Gives guests' orders to the chef and other kitchen employees accurately and in a timely manner
	Expedites orders and verifies that they are correct and complete before serving guests
	Serves all menu items to customers in a timely and courteous manner, according to the management's dictates
	Provides continuous service to customers throughout the meal
	Totals customer checks with 100 percent accuracy and presents checks to customers
	Thanks all customers for their patronage and invites them back

In food service there will be a culinary staff, a server staff, a managing staff, and perhaps others. Within each staff will be various positions. For example, within the service staff the positions are a host or hostess, servers, buspersons and others. Each staff works together as a team to perform its functions, and the various staffs then blend their efforts together to achieve their goal. All servers should realize this and make themselves a part of the team. If they do, it will make their job a better one. Individuals working alone are without help. Being a member of a team helps to carry one along much more easily and smoothly.

Leading

An operation cannot succeed without effective leadership. To be a good leader, one must know what the goal of the operation is and then take firm and adequate steps to reach that goal. Good leaders must command respect and good will, as well as success and profits. Good leadership motivates workers to do their best, and sees that workers are treated fairly and are given a chance to reach their goals.

Leadership styles run the gamut, from strict authoritarians to passive leaders. The recommended style today is a mixture of these, called **participative leadership.** In this style of leadership, the manager is a part of a team working with the employees toward a common goal. Traditionally, hospitality industry leaders have been authoritative, or dictatorial. There are reasons for this: (1) The nature of the industry was extremely competitive, with little time to promote a team atmosphere; (2) Managers had little time to train employees and had to set up tight control measures to see they were followed; (3) Because of the high employee turnover in the industry, managers were not able to build teams among their staff. As the labor market and the industry changed, managers found they had to change their management style to survive.

While authoritative leaders rule with little or no employee input and use fear of discharge or punishment to motivate action, participative leaders act as coaches to lead teams to success. Such leaders seek out ideas and opinions of workers and often follow them, giving employees a feeling of value and respect. They work alongside employees when necessary, creating a feeling of fellowship and the sharing of responsibilities.

Participative leaders still must ensure that standards are met and that rules are applied consistently and fairly. Participative leaders, rather than reprimanding and punishing, point out problems to employees and help correct them. This is why standards are so important, so that employees understand what is expected. It is important to specify the corrective action when pointing out mistakes, and to be as positive as is appropriate. Managers should correct the words, actions, or attitudes, not the employee personally. Counseling sessions should be held in private without any tones of anger or threats. They should be done more to make the employee feel they are being offered help rather than criticized.

Controlling

Controls lead an operation toward achieving its goals. Without controls, operations lose money and workers become frustrated and unhappy. Establishing standards and procedures for meeting them is key. Examples of **service controls** include the following:

- Greeting guests within one minute of being seated
- Serving lunches in less than fifteen minutes
- Receiving fewer than ten guest complaints per month
- Knowing the specific steps to take when guests complain

Servers and managers should check continuously to see that deviations from standards are avoided or corrected.

Motivating Servers

Management is responsible for seeing that employees are motivated and that they work in a motivating environment. To ensure good service, managers must develop a staff of servers that want to do the best possible job and do it consistently. Developing a motivated staff is not easy and management should make it one of its most important goals.

People are motivated to work by a number of factors. Primarily are the basic needs for food, housing, transportation, and so forth. Next comes the need to work in an environment that is safe and secure. Following this are social and environmental needs; people want to feel in any job situation that they belong and they are an accepted member of a group. A number of people also are motivated to work just to get a change in environment—elderly people may work because they are lonesome and bored with their home surroundings. Next comes the need to satisfy self or ego needs—people may like the independence of having their own income or just want the pride of having a job. And lastly, one may work to realize self-actualization or achievement needs—a teenager may become a busperson not only to have an income but to assert maturity.

It is a mistake to think that everyone is motivated in their work by all these needs. Some workers barely rise above basic security and safety needs, while others are motivated by all of them. Teenagers have different needs than more mature adults, and elderly workers often have different needs than those in other age groups.

Managers should not only try to satisfy the needs that workers have, but should also seek to extend them to higher levels. It will result in more satisfied employees. Every individual also has pet needs. If management can satisfy these, a more satisfied worker is made.

Hiring Motivated Servers

As indicated previously, hiring is the first step to achieving motivated employees. Often the motivated employee can be detected in the interview by certain personality traits:

- A genuine interest and respect for people
- A warm and outgoing attitude toward life
- A sincere wish for the job
- Habits that will not interfere with the job
- A willingness to accept an entry-level position if they have little or no experience
- A positive job record
- Poise and confidence
- An ability to solve problems and make good decisions
- A pleasing appearance
- Experience working in good operations, if applicable
- Positive reasons for working other than income
- Willingness to grow in the job and eventually be promotable
- Good self-direction and self-reliance

People must possess self-motivation to work hard and serve people enthusiastically. The following interview questions might help in evaluating an applicant's capability of becoming a well-motivated server.

1. What was the biggest challenge you had in your current/last position? How did you meet that challenge?
2. What are your strengths? What are your weaknesses?
3. Have you ever had problems following instructions?

4. What type of goals have you set for yourself?
5. What do you see yourself doing five years from now?
6. Do you consider yourself a person that performs service duties with enthusiasm and dedication?
7. Do you like to deal with and be around people?
8. What do you expect from your coworkers?
9. What do you expect from your supervisors?
10. Can you point out how you introduced a new idea that improved your work or the work of others? How did it make you feel?
11. How would you handle a guest who is irate because the order was served late and the food was cold?

There are laws that restrict what one can ask while interviewing people for jobs. These relate to questions about things that are not job related. For example:

1. Have you ever been arrested or accused of stealing?
2. Do you have a green card?
3. How long have you lived at this address?
4. Are you the wage earner in your family?
5. Are you married? What does your spouse do?
6. How many children do you have or plan to have?
7. How old are you?
8. What would you say your credit rating is?

Creating a Motivating Environment

One of the most important things managers must do to motivate employees is to see that a respectful, pleasant environment prevails. This helps servers feel both secure and appreciated, as well as motivated to give good service. One highly successful manager has said, "I treat my employees as I treat my customers; both are valued." A positive environment fosters team work and cooperation among servers. This is essential in the often pressure-filled, fast-paced, and demanding profession of serving.

To ensure good service, one must develop a staff of servers who want to do the best possible job, and do so consistently. One must create a feeling of respect and commitment between team members, with managers participating in this team. Establishing standards and enhancing jobs such as empowering workers, or doing all one can to make the work easier are good motivators. Encouraging open communication, including servers in goal setting, treating employees fairly and with respect, and creating awards for good performance are others.

Some managers do things that destroy motivation:

- Assigning extra work without adequate compensation in praise, promotion, or other recognition.

- Demonstrating unfair or nonuniform treatment or aggressive, abusive, cold, inconsistent, or stand-offish behavior.
- Failing to provide good communication, supplies, tools, or other standard conditions.
- Ridiculing or using sarcasm instead of constructive criticism.
- Oversupervising and showing a lack of independence and trust.
- Failing to follow up on problems mentioned by servers.
- Being indecisive or unwilling to support servers.
- Making decisions, without consulting servers, that might interfere with their being able to adequately do their jobs.
- Having too many bosses.

Empowering Employees

Empowerment is a management tool in which employees are given the power to make decisions that achieve higher standards of service. This power includes the authority and responsibility to make things happen. For example, instead of a server having to go to a superior to approve a meal deduction, the server is empowered to make the decision in order to correct the problem. Servers may be given some guidelines so they know what is and is not appropriate. Empowerment benefits servers, who are likely to feel more personal commitment to serving guests well. Empowered employees likewise benefit employers. Problems solved more quickly are better served. Dissatisfied customers are helped more quickly, and situations are resolved.

Rewards as Motivators

There are many ways management can let employees know when they are doing a good job and thus motivate them toward doing an even better one. Positive feedback and recognition of a job well done are very valuable. Monetary rewards and prizes may also be given. Many operations have an employee-of-the-month and employee-of-the-year award. Recognition of servers to guests is very powerful. Some operations offer servers bonuses after a certain length of employment. Others use extra vacation days, dining credits, concert tickets, amusement park passes, and trips to reward excellence performance. Achievement is traditionally rewarded with advancement.

Rewards should be open to all in the same category, and all must be given complete information on what is required to get the reward. Monitor the program and evaluate it; does the program achieve its desired objective? Give the award as soon as possible after it has been won and tell them exactly why they qualify. Make the award worthwhile. The award program should be respected by the servers.

Things like offering adequate health care for servers and their dependents, giving educational assistance or assistance in personal problems, and even day care assistance for children can be good motivators to keep servers on the job. Whatever the incentive, managers should give them consistently and tie them to concrete behaviors. Special favors will undermine any incentive plan.

Remember that respect and loyalty cannot be purchased. It must be earned through sincere actions. Servers will see through attempts to be manipulated. Unless incentives are given appropriately, they will act as demotivators rather than motivators. Money alone rarely satisfies employees. In a survey taken of what employees want, the following motivators were mentioned frequently:

- Job security
- Good working conditions
- Appreciation
- Job satisfaction
- Good wages
- Acceptance
- Dignity and respect
- Good benefits

Note that only two (good wages, benefits) of the eight items are monetary; the other six deal with psychological factors that can cost little or nothing.

Unless servers get a sense of gratification from doing a good job, they lose interest. The work must have meaning to them and it must help them reach their own personal goals. Managers need to know that they cannot reach their goals unless employees reach theirs.

FIGURE 10.1

Rewards are great motivators for employees and can help promote job satisfaction. Courtesy Corbis Digital Stock

Evaluations as Motivators

The formal way of giving employees feedback on how well they are doing is the performance evaluation, or appraisal. Managers should make it clear that the purpose of evaluations is to help the employees develop and improve. It is important that servers know that evaluations are positive, not negative. Evaluations should let employees know their opportunities for advancement.

The evaluator should look at customer comments, dollars produced, tips, covers served, and other factors to make the evaluation. Other intangible factors, such as attitude, teamwork, and punctuality, should also be included. A performance record should be kept.

Appraisals can be helpful to management in giving information on employees attitudes, goals, needs, and concerns. The information gained can also be helpful in giving advancement, pay increases, rewards, and what management must do to further the employee's pursuit of goals.

A good appraisal leaves employees more sure about how well they are doing and what has to be done to improve. The things that need to be done for advancement, better pay, or other benefits can be learned.

Scheduling Servers

Proper scheduling can be a good motivating factor. It is essential that all workers be treated fairly in scheduling. Some operations have a scheduling system that rotates servers among the different stations because some stations are easier to work or give better tips than others. Requests for special days off or certain hours should be given consideration and allowed, if possible.

Managers are responsible for seeing that adequate, competent servers are scheduled at the right times. If too little help is on the floor, good service is impossible. If underscheduling is consistently done, servers will slow down and management will create a very difficult problem for itself (see **Exhibit 10.4**).

The number of covers in a station should be assigned according to the stations' distance from the kitchen. Team service allows at least one server per station to attend to guests while others do other work. Experienced buspersons also help in placing orders and delivering food and beverages.

Managers should take advantage of computer software programs that can help in planning better station assignments, scheduling, analyzing payroll, and monitoring service.

EXHIBIT 10.4 Sample Server Schedule

	Monday	Tuesday	Wednesday	Thursday	Friday	Saturday	Sunday
Jenn	Off	10 a.m.–3 p.m.	Off	10 a.m.–4 p.m.	10 a.m.–3 p.m.	10 a.m.–4:00 p.m.	11 a.m.–7 p.m.
Terry	Off	11 a.m.–7 p.m.	10 a.m.–4 p.m.	10 a.m.–3 p.m.	Off	11 a.m.–7 p.m.	10 a.m.–3 p.m.
Damon	Off	10 a.m.–4 p.m.	10 a.m.–3 p.m.	Off	10 a.m.–4 p.m.	10 a.m.–3 p.m.	10 a.m.–4 p.m.
Kate	Off	Off	4 p.m.–close	3 p.m.–close	3 p.m.–close	4 p.m.–close	3 p.m.–close
Carlos	Off	3 p.m.–close	11 a.m.–7 p.m.	4 p.m.–close	11 a.m.–7 p.m.	Off	4 p.m.–close
Lisa	Off	4 p.m.–close	3 p.m.–close	11 a.m.–7 p.m.	4 p.m.–close	3 p.m.–close	Off
Deb	Off	3 p.m.–10 p.m. →					→ Off
Kyiel	Off	Off	4 p.m.–close	10 a.m.–4 p.m.	4 p.m.–9 p.m.	4 p.m.–9 p.m.	4 p.m.–9 p.m.
Shawn	Off	11 a.m.–7 p.m.	10 a.m.–3 p.m.	6 a.m.–11 a.m.	3 p.m.–close	Off	5 p.m.–close
Rosa	Off	10 a.m.–4 p.m.	Off	2 p.m.–9 p.m.	10 a.m.–4 p.m.	11 a.m.–6 p.m.	4 p.m.–10 p.m.
Dan	Off	4 p.m.–close	10 a.m.–3 p.m.	Off	11 a.m.–6 p.m.	5 p.m.–close	11 a.m.–6 p.m.
Yi	Off	2 p.m.–8 p.m.	10 a.m.–4 p.m.	4 p.m.–close	Off	5 p.m.–close	10 a.m.–3 p.m.

Dining Room Arrangement

The dining area must be planned properly to allow for good service. Aisles through which servers must move while carrying heavy trays must be wide enough to allow for safe and good passage. Often managers are tempted to add too many tables, cutting down on space for servers. This is one good way to achieve poor service. Twelve square feet per guest is the minimum for regular table service and 20 square feet per guest is required for counter service. Club and luxury dining areas often have more than 12 square feet per guest. Banquet service requires less. Distances between tables should be 4 to 5 feet so that aisle space between servers is not restricted. The table should allow at least 24 inches of linear space per cover. Thus, a table for four should be at least 30 inches square.

Managers should see there is adequate room for tray stands, service stations, and other service equipment. One to four servers typically use one service station; the fewer the better. If possible, water should be piped to the station. In fine-dining operations, a carving station may be placed among tables for carving, deboning, and other activities. Mobile service or carving stations may be used. These can save space and give more flexibility in arranging tables.

Tray stands, service stations, and other things needed to give good service should be kept in order and be sufficient in number.

Kitchen Arrangement

Managers will assist servers by seeing that a smooth-flowing kitchen arrangement is set up. Unnecessary backtracking leads to delays, accidents, and frustration. A melee of workers going in all directions during busy periods makes things hectic for everyone, including cooks. A smooth, one-way flow, with pick-up arranged in sequence of courses, is best. This might be difficult to achieve since one section may be responsible for several courses. (A pantry might serve cold appetizers and salads while the grill area prepares only grilled items.) In these operations, sections should be placed apart from each other so as not to interfere with flow. Mobile equipment can be used to move items between stations, reducing server travel.

It is essential that food be delivered to guests with the proper appearance and temperature. Guests should not have to wait long for food. Efficient traffic patterns help ensure satisfied guests.

Training Servers

A wise manager would never send an untrained, inexperienced server out to work a station alone. Servers must be trained in what to do, how it is to be done, and the professional standard expected. Although managers are responsible for seeing that servers are properly trained, the training function can be delegated to others who can do a good training job, who themselves must be taught to give good training.

Although service training can be costly, poor service is more costly in terms of lost guests and profits. Here are some of the benefits of good training:

- Improved service and increases number of satisfied guests
- Improved productivity, reducing both employee turnover and labor costs
- Reduced waste, accidents, and breakage
- Opportunities for skilled, knowledgeable, confident servers to grow and advance
- Lower frustrations
- Reduced turnover

Training should be an ongoing program in all operations, because one never stops learning. Training should improve knowledge, skills, and attitudes.

The Effective Trainer

A good trainer knows people, how to teach them, and how to motivate them to learn. Good learning can only take place when the trainer knows what must be taught and can teach this to the trainees. Before one starts to teach, one should have a lesson plan. This helps the trainer to stay on the objective and not stray, helps keep the program on time, and organizes the training session. Trainers need to be good leaders and must be able to control the learning session. They should be adept in building a feeling of trust and confidence with those they teach. A good trainer is a good listener as well as speaker. Students should be stimulated to ask questions and become involved in the training session. The learning given should have meaning to the student. Besides indicating the what, how, and when for jobs, the why should also be given. A student knowing why is more apt to make the learning permanent than just hearing the what, how, and when. Trainers should not be afraid to repeat; repetition reinforces learning. They should also seek feedback on how well the student has learned.

Trainers should know the jobs they teach thoroughly and be able to communicate such knowledge to others. Not everyone can be a good teacher. One must have a personality for it and like to do it. Persons who are to train others should be trained for it. Although teaching may come naturally, one can improve on that natural ability by knowing some of the techniques of teaching.

The trainer should plan detailed training and orientation sessions. Giving a pretest enables trainers to identify in what areas employees need the most training. Never assume that a server, even those with extensive experience, knows everything. It is a good idea to test servers after a training session, and then periodically test following the session. This ensures application of training on the job.

The Training Session

Lecturing, group discussion, role-playing, show-and-tell, and on-the-job training (OJT) are some of the ways used to train servers.

Lecturing is good when the material to be covered is short and not too technical. It is good for imparting broad and overall information. Any lecture session should be short. Listeners quickly tire of just listening. It is best to use lecturing with the other techniques of teaching.

In planning the format and preparing the material for a learning session, the trainer will greatly benefit by keeping in mind that trainees remember:

15% of what they hear
30% of what they see
50% of what they hear and see
85% of what they practice

Group discussions provide motivation, interest, and subject retention for those being trained. They can be stimulating and very beneficial in getting learner interaction. The trainer must first identify the discussion objective, relate the topic to the learning objective, and manage the discussion. Trainers should try to get the discussion started and then stay out of it, only helping from time to time to direct the discussion.

Role-play allows servers to see and practice what is to be learned. It gives servers a chance to review and criticize, and helps to change or strengthen attitudes, skills, or knowledge. The active participation teaches one to work with others. The first step in role play is to give the trainee a situation relating to their position, and then allow the trainee to work his/her way through it. Role-playing is best used where active, physical effort is needed, coupled with the demonstration of some skill and knowledge. The proper way to perform may be explained by the trainer and then the student or students allowed to go perform the task. A critique should follow. Effective learning occurs if the others in the class join in the critique.

A **show-and-tell** method gives students a chance to hear and see how something is done, perform it, and then see where improvement is needed from the trainer and perhaps others in the class. Such feedback reinforces learning. Show-and-tell usually is used in classroom situations.

Usually there are four steps in show-and-tell:

1. Tell the server how to do the task.
2. Show what is to be done, how to do it, and why.
3. Have the server repeat the task, verbally explaining what is done and why.
4. Have the trainer and/or the class review the performance.

Observation and on-the-job training (OJT) usually occurs in the workplace under real conditions. It can be quite successful if done correctly. Observation allows trainees to follow the trainer through the tasks so they can see how they are correctly performed, and then under a real situation go through the same tasks. The trainer observes and later provides feedback in private. OJT must be done carefully so as not to lower the quality of service to guests. Trainers should be competent to train and know the correct ways work is to be done. Too often the trainee learns, but learns the wrong way because the trainer did not know how to do it.

FIGURE 10.2

On-the-job training is a valuable method for instructing new employees. Courtesy Action Systems, Inc.

The Training Space, Tools, and Equipment

Some operations have wisely established their own training manuals. These usually cover such areas as greeting guests, giving menus, taking orders, and sequence of service. Mise en place (work done to get ready for guests' arrival) information should cover linen arrangement, table setup, sanitation and safety practices, and stacking service stations. Menus and preparation that servers need to complete orders should be included. Selling techniques should be discussed and employees trained to use them. **Job descriptions** should indicate servers' tasks and responsibilities. They are good learning materials because they describe all the tasks done.

Externally produced training booklets, videos, software packages, CD-ROMs, and posters are available from professional associations and publishers. They are usually of high quality, having been prepared by authorities on various service topics.

Training models and formats vary according to the size and type of operation. Among the most common are: programmed learning by exercises, on site demonstrations, classroom activities, co-op programs, apprenticeships, and on-the-job-training.

Reservations

Management is responsible for establishing the reservation system and seeing that it operates as desired. Often employees operate the reservation system and sometimes the employee may be a server.

The reservation taker should show warmth and cordiality in taking the reservation. This person provides the first impression of the operation, and this is important. Follow the greeting with the name of the operation, perhaps adding, "How may I help you?" The conversation should be short and precise. Check the reservation book before confirming a reservation, noting the date, time, and party number. At the end of taking any reservation, the taker should thank the caller.

Some reservation takers tell guests to please inform them if plans change. No-shows can create problems. It is advisable to ask for a number where one can reach the guest, especially in the case of a large party. This number may be called on the day of the reservation to be sure the party is still planning to come. Some operations take a credit card number and may charge for a no-show. Sometimes a request for a reservation is made too far in advance. A note can be made of this, and the guest can call at a later date. The seating preference should be noted, especially smoking or non-smoking, if the restaurant has a smoking section. Other things to note on the reservation book are occasion, special menu request, and any special needs.

It is important that reservations be arranged to get the best turnover possible. It is usual to stagger reservations to avoid having everyone come in at one time. This avoids overloading kitchen and service staffs. Some operations refuse to take reservations because of no-shows, and cancellations. They usually have enough walk-in business to make the operation successful, and a reservations system would be a burden rather than a help.

To establish a reservation system, first determine the number of tables and the number of seats available, and if tables will be moved to include different sized groups. From this, set up a reservation chart. (See **Exhibit 10.5.**) Usually a table is filled at dinner time for 1½ hours for a table of one or two, but for larger groups the time is longer. Of, course, the meal has much to do with it. Breakfasts have a fast turnover, lunch has a little longer, and dinner has the longest stay. However, a breakfast group having a meeting, or a meeting for some other purpose, may hold for two or more hours. An understanding about how long tables will be filled will give the reservation person an idea of the timing of reservations.

EXHIBIT 10.5 Sample Reservation Chart

A reservation chart details where different tables and sections are located in the dining room.

LUNCH RESERVATIONS FOR JOHN PURDUE ROOM

Name	Number in party	Phone	Your initials	Smoking/ no-smoking	Special requests
11:30					

Name	Number in party	Phone	Your initials	Smoking/ no-smoking	Special requests
11:45					

It is not unusual to overbook by about 10 percent of the dining room capacity. This is done to take care of no-shows, cancellations, and a slow walk-in night. This can, at times, lead to problems of having too many guests arrive with no tables to receive them. When this happens, the following is suggested:

1. Be consistent in handling situations.
2. Ask the party to please wait a few minutes, saying they will be seated as soon as possible. Telling a little white lie that the guests at the table are staying a little longer than expected is all right. If the wait is long, perhaps invite the group into the lounge for a before-dinner beverage. Or, say you are sorry, but to make up for the delay you will serve the party a complimentary bottle of wine.
3. Give the time of wait if possible, but be sure the time is as accurate as you can make it. Do not say, "It will be just a moment," if you know it will be longer. Nothing causes more frustration than being given a time and then having it be far off. A slight difference is not important. If one cannot give an accurate time, tell the party, "Your table will be ready as soon as the party seated there leaves."
4. When approaching several groups waiting to be seated, pleasantly tell the group or groups whose tables are ready, and then reassure the other groups that they have not been forgotten.
5. A seat card prepared for the specific needs of the establishment can be a very helpful tool. Upon seating the guests, the host, hostess, or maitre d' will give the card to the server. The card contains all the information the server needs to know in order to follow up during the course of the meal: name of the host, table number, time seated, eventual time frame, special celebration, special requests, dietary needs, and so on.

   A typical seating card might read:

   Mr. and Mrs. H. Johnson–2

   Time in: 7:30 p.m.

   Table #8

   Server: David

   Notes: Anniversary cake, Mrs. J. allergic to seafood

There are many other things management must do to see that good service occurs. Often, these duties do not directly influence service, but do so in an indirect way.

A manager is responsible for seeing that good employee records are maintained, that tips are paid to employees and recorded, that vacation time is appropriately planned, and that employee activity programs, such as the establishment of a ball team, a dance, or other events are fully encouraged and supported. Management should work to make the whole operation one cohesive family. It not only leads to more motivated employees, but to the profit of the operation.

CHAPTER SUMMARY

Managers are responsible for good service and should help their employees achieve it. One of the things that can help develop good service is a well-run operation. A poorly run operation frustrates employees. To run well, a management must properly practice the five functions of management—planning, organizing, staffing, leading, and controlling.

It is management's job to see that servers are motivated to achieve high standards of service. A part of this comes in the hiring, but management has available many motivating rewards it can use, and many of these cost nothing. A warm environment, and fair and helpful treatment can motivate employees. Management should be a member of a team that wants to achieve. Participative management is usually the most successful.

Management is responsible for proper scheduling so servers know where to be when needed. This is management's job. Management is also responsible for seeing that the dining room and kitchen arrangements are set up to encourage good service.

Training is essential for good service, even with experienced servers. Training programs should be set up so servers know their work and possess the skills required. Training is management's responsibility and should be ongoing.

Complaints are often handled by servers, but sometimes management must step in and handle difficult ones. Knowing how to handle complaints is essential to prevent dissatisfied customers. Complaints can often be avoided by recognizing when the potential for one exists. For this reason, it is strongly recommended that managers and supervisors spend time with the service staff discussing how anticipating customer needs leads to better service.

Reservation programs are also the responsibility of management but at their establishment others may do the reservation taking, and sometimes these people are servers. The reservation taker should get complete information such as date, time, number in party, and other information desired by the operation. Some operations refuse to take reservations because of no shows, cancellations, and other problems relating to a reservation system. The reservation taker should show warmth in taking the reservation and should thank the caller when the reservation taking is finished.

RELATED INTERNET SITES

Bureau of Labor Statistics

The occupational outlook for foodservice managers provides an insightful discussion about the intricacies of the food service industry, as well as expected salaries for various segments.

www.bls.gov/occ

KEY TERMS

delegation
empowerment
five management functions
job descriptions
line organization
participative leadership
reverse pyramid
role-play
service controls
show-and-tell
standards

CHAPTER REVIEW

1. Why are standards important to good service? Who should establish them?
2. Describe line organization. What are some advantages? What other type of organization is used today instead?
3. What are the characteristics of a participative leader?
4. Create a list of questions you would ask a person applying for a job. Are the questions lawful? That is, do they violate a person's privacy?
5. Name several faults in dining room arrangement that can cause problems for servers. Name some in kitchen arrangement.
6. What are some benefits of having a good server training program?
7. What information should one get from a guest in making a reservation?
8. What should the host or hostess do if two parties arrive at the same time, with reservations, and there is only one table available?
9. One server is so busy with a large party that she hasn't been able to approach a table that was seated five minutes ago to take the order. What can the other servers do to help?
10. There are a number of duties that managers must perform to ensure that good service occurs. Name some of these.

Case Studies

Meeting Competition

Ralph Martin has, for a number of years, operated a successful family restaurant in a large shopping center. However, he finds that in the last few years new operations have come into the area and reduced his business. If the trend continues, he will soon be operating a losing restaurant.

Two of his competitors are fast-food operations. Another is a family restaurant like his, but with newer decor and menu. This operation serves a quicker food menu with more emphasis toward meeting guests' nutritional concerns. This family food competitor recently introduced an early-bird menu to attract budget-minded customers and the elderly who like slightly smaller servings at lower prices and enjoy eating earlier.

Martin wonders what his options are. During his years of operating his restaurant, he has been able to save enough to retire. Should he look for a buyer, sell out, and retire? Despite his savings, he feels that he has a few good years of active business life left in him, and that he will miss the busy life of running a restaurant. He is rather doubtful that he would like retired life right now. He would like to set aside a little larger nest egg, to afford more travel when he is retired. In addition to everything else, his competitive nature is not happy letting another business best him in a game in which he has been successful for years.

What should he do? Should he revise his menu? Are there any ways you can think of meeting the competition that would make his operation stand out as different from others? His staff belongs to a union, and over the years he has done so well that he has let his employees' wages get higher then the norm. Should he call his workers and the union together to discuss the problem? Should he contemplate the renovation of his facilities and introduce a more modern menu? What's the risk of spending so much money in renovating? Might it fail to do the trick?

Write up your assessment of where you think his main problems are and how he can try to solve them. Can you find adequate and good reasons for bowing out that might be of greater weight than those given for his staying in business?

Projecting Costs for a New Operation

Jonathan Olinas, a successful operator of cafeterias, feels that there is a good opportunity to start a cafeteria in a thriving shopping center in a newly established business district. It will cater to shoppers and families. The income level of the families surrounding the shopping center is above average.

He estimates probable revenue and is pleased with what he sees as income by estimating numbers of shoppers and estimating probable income from that source based on demographics. He does the same in estimating family patronage

and income from that source. There is a probable source of income from take-out, which he estimates as well. He is encouraged with what he finds.

He then does a study and comes up with the following estimate of costs:

Food and beverage cost	38.1%
Payroll and benefit costs	39.0
Operating expenses	5.8
Advertising and marketing	2.2
Repairs and maintenance	1.5
Rent	6.1
Taxes	0.6
Insurance	0.7
Depreciation	2.5
Administrative and general	3.8
Miscellaneous expenses	2.8
	103.1%

Of course he is disappointed. He wants to make a 4 percent to 5 percent profit.

What would you advise him to do? Give up the idea, or make changes that will bring costs into line? Reducing the costs by 7 percent to 8 percent is a challenge. Make suggestions for decreasing expenses and then describe how you would go about implementing the changes. Your changes should be realistic, backed with possible ways of how he is going to do it.

Hiring

Antonia Smith is the operator of three upscale restaurants. She needs servers who are above average in their skills, are able to please guests, are highly presentable in their dress and appearance, and can learn how to serve guests of upper-scale dining. Tips are good and a number of her wait staff make a satisfactory living by just working as servers. She hires many college or university students. She finds they are bright, flexible, and quickly fit into her system of service. However, it is difficult to fill her server needs from this source, and she finds she is being forced to take servers who do not come up to the standards she believes she needs for her type of operation.

A person in her office is assigned to interview and select staff. This person is well educated and very capable of selecting a high-class employee, the kind she needs. However, in several instances this interviewer got her into trouble with authorities because of questions asked in interviewing applicants for jobs.

Set up a guide for this interviewer to follow in interviewing. Also, describe how can she attract more servers who are college or university students. Would putting a student representative on the campuses be a good idea? What kind of advertising might be effective? Would a cash reward for bringing in employable candidates be helpful? Address these questions and describe other ways you think Antonia Smith could recruit the staff she needs.

TABLE ETIQUETTE

OUTLINE

I. Introduction

II. A History of Table Etiquette

III. Principles of Public Dining Etiquette

- A. Pre-Service Etiquette
- B. Tableware
- C. Etiquette for Specific Foods
  1. Finger Foods
  2. Soups
  3. Breads
  4. Vegetables
  5. Salads
  6. Pasta
  7. Fish and Shellfish
  8. Desserts

IV. Tipping

V. Dining Etiquette of Various Cultures

- A Traditional Jewish Service
- B European Dining Etiquette
- C. Chinese and Japanese Dining
  1. Chinese Dining
  2. Japanese Dining
- D. Indian Dining
- E. Middle Eastern Dining

VI. Chapter Summary

VII. Chapter Review

VIII. Case Studies

LEARNING OBJECTIVE

After reading this module, you should be able to:

- Describe the etiquette rules concerning special foods.

Introduction

The way customers are served food and drink often greatly influences how they proceed in eating and drinking it. The proper items for their consumption must accompany them, and placement and other factors must be correct. Because of this, servers should know the basics of service and dining etiquette so guests can enjoy their meal.

The failure of a server to serve properly may cause a guest to feel uncomfortable and not enjoy a meal. Serving properly includes not telling a guest how to eat and drink. The adage, "To possess good manners is to not tell others what good manners are" holds true. Saying to a guest, "I brought you a bowl for your clam shells," lets the guest know what to do without embarrassment. If a guest looks uncomfortable, astute servers can help lead the guest through the meal. It is every server's duty to help prevent the guests from feeling embarrassed or uncomfortable because of etiquette, and every manager's duty to see that servers are trained in proper etiquette and service.

A History of Table Etiquette

People have gathered to dine together for thousands of years. Undoubtedly, certain rules of decorum arose in various cultures suited to the needs of that culture. However, we find no written record of these rules until near the end of the Middle Ages (454 A.D.–1474 A.D.) in Europe. In 1474, a treatise titled *De honesta voluptate et valetudine,* "Health and enjoyment within decency," written by Platina da Cremona (whose actual name was Bartolomeo de Sacchi) appeared. It proved to be a popular work on the art of good living and manners. In 1507, a manual titled *II Cortegiano, The Courtier,* by Baldassare Castiglione, dealt largely with proper conduct in public but also covered dining. It immediately became the accepted authority. Even some educational institutes used it as a text for teaching ethical behavior and morality. Although *The Courtier* won wide acclaim with aristocrats and educators, the most popular work on the subject appeared in 1555, a work titled ***Galateo*** by Giovanni Della Casa. This short, unpretentious work quickly became the authority on good manners, and in a few years was translated into every language in Europe. The author delivers his message in an imaginary conversation between an old wise man and his young nephew who is getting ready to confront the realities of life. The brief narration centers on the teachings of this old master, who is concerned about eventual social errors the youth might commit. The advice is specific and practical:

> While seated at the table it is against good rules to scratch oneself. A person has to prevent the practice of spitting and if it has to be done, then it must be done in a decorous manner. I have heard of nations where people are so well behaved that [they] never spit. We should avoid taking food with avidity; it will help avoid hiccups or any other unpleasant sequence. It is ill-mannered to

rub a napkin or a finger against the teeth and to wash the mouth with wine and spit. It is also improper to take small items like toothpicks from the table and hold them in the mouth looking like the beak of a bird that is building a nest, or holding them over the ear like a barber. Some gentlemen seat on [sic] the table already with toothpicks placed in the clothing around their neck. Going that far I wonder if they should come to the table with spoons tied around their neck!! It is incorrect to lean all over the table and eat so much food in one time that both sides of the mouth and both cheeks look swollen. . . .

Other early notable works on table etiquette are **A Treatise on Manners** written by Erasmus in 1526, and a collection of the principles on proper table etiquette by Robert de Blois, which appeared in the middle of the sixteenth century. In 1765 Antoine Le Courtin authored the **Traite de Civilite** in which he summarized all of the previous writing and offered some advice of his own according to the needs of the time. There are also interesting accounts in Japanese, Chinese, and Arabic literature on the rules of good behavior when eating. Many of the customs in these cultures differ greatly from the Western cultures.

Principles of Public Dining Etiquette

Our rules for dining etiquette grew out of what is practical, makes for good social functioning, considering others, accommodating people in social situations, and facilitates the art of eating and dining. They were not formed by one authority but evolved from common usage of dining practices and social relations. Authorities have collected, summarized, and published them, but people themselves formed them.

Rules of **etiquette** are not set; they change constantly as society changes. Today's rules are less formal than they were during England's Victorian era. There is also increased awareness that customs and etiquette differ from culture to culture. Rules differ from occasion, time, place, and company. Two senators might treat each other quite differently on the U.S. Senate floor than they would on the street. Today, we have accepted norms for eating out in public, but they are not rules that cannot be violated. Servers should learn the rules and follow them when appropriate, while realizing that all rules depend on the situation at hand.

FIGURE 11.1
Children can learn proper table etiquette by following adult examples. Courtesy PhotoDisc, Inc.

Pre-Service Etiquette

Some food service operations take reservations. In general, the employee taking the reservation should get the party's name, request for a smoking or nonsmoking section if appropriate, number in the party, arrival time and date, and any special arrangements or requests. When guests with reservations arrive, if there is a wait, apology for it, say how long the wait will be, and where they should wait. Some operations won't seat a party until all of the party is present. Others will hold a reservation for only 15 minutes. Some operations call the person who made the reservation on the day it is to be honored to remind the guest of the reservation time. There are operations that make a charge against a guest's credit card for a no show. Coats should be checked so that the table and seats are clear for the guest. The maitre d', host, or hostess should ask guests for their seating preference before leading them to a table. If a preferred table is unavailable, tell the guests how long a wait will be and what their options are.

When guests have seated themselves, the server may place napkins on guests' laps or allow guests to do this themselves. It is considered improper for guests to tuck the napkin under the chin or in any piece of clothing, although it can be discreetly fastened under a belt to prevent the napkin from falling. Do not give the napkin a violent shake to open it. Napkins should be placed on the table so guests can pick them up by the closest edge and unfold them easily.

One of the first things the server will do is ask the guests for their drink order. It is considered impolite to urge guests to drink an alcohol beverage if a guest prefers not to drink alcohol. The server should suggest a nonalcoholic beverage.

Servers should announce specials and their prices clearly and slowly. Giving some time between specials helps guests evaluate each one. The delay, however, should not be too long.

Sometimes guests wish to share each other's orders. Let guests know that they should ask the server to have this done in the kitchen and avoid doing it at the table. Guests may desire to taste someone else's food. The person with the food can place a small amount on a plate passed by the other person. Or, the person with the food may pass over the whole plate while the other takes what is desired and passes the plate back. Take-home containers are acceptable except at formal meals and banquets.

It is considered rude for any guest to whistle, clap, snap fingers, strike a glass or cup, stamp feet, or make other noises to attract a server. Turning toward the server and raising one's hand should be a sufficient gesture to indicate the server's attention is desired. If the server is close by, one also might call, "server," in a soft voice to get the server's attention.

Sometimes a guest is served something that one must avoid because of an allergy or dislike. The proper thing for a guest to do is to not mention the problem but to avoid the item and eat the others. If something must be removed from the mouth, it can be brought to the front gently and put onto a fork or spoon and then placed down on a plate. It is also all right to move the item from the lips with the fingers, such as a fish bone. Inedible items should not be spit out, but also removed as unobtrusively as possible and put onto a plate. If a guest sees something inedible on any food, it should be called quietly to the attention of the server to be removed. If an item is dropped, one should use a fork or spoon to pick it up and then put it on a plate to be removed with the plate. A large spill requires the attention of the server, who must clean it up and then cover it with a clean napkin.

Guests should not reach, but should ask to have things passed. It is helpful if a guest holds a dish for another while the other is helping herself or himself from it. When taking something solid from a dish, it is helpful if one can use the **Russian technique,** using the spoon and fork to lift and carry the item to the plate. In passing items with a handle, pass with the handle pointed toward the person receiving the item.

Butter and other items often are served in small containers. Guests should remove the cover, take pats of butter with the knife or fork provided, place on the butter plate, and replace the cover. The same is done with jellies, jams, and other solid or semisolid items. Servers should remove and replace empty containers as soon as possible.

It is proper to wait until all have been served before eating, but in large groups this could mean that some food will get cold. In this case, one should eat when served. Often the guest not yet served will try as much as possible to urge others to eat as soon as served. Servers should serve all guests in the party at the same time. Servers might wait until all guests are finished with one course before clearing soiled dishes. However, it is not improper to clear one guest's plate, while another is still eating. Some guests do not like to sit with their soiled dishes in front of them. If in doubt, the server should always ask, "May I take your plate?"

Tableware

Especially in formal dining, eating utensils are given precise placement on the table for specific uses. Knives and spoons are placed to the right of the guest, while forks are to the left. In the United States people cut their food then transfer the fork from the left to the right hand, a process sometimes described as **zig-zag method.** Europeans do not use this zig-zag method. After cutting they leave the fork in the left hand to bring the food to the mouth.

The fork should be held at a 40-degree angle with the tines pointing down, holding the food to be cut, with the thumb under and the second finger giving support on the right side of the handle. The index finger should be placed on top of the back of the fork's handle, pushing down. It is acceptable to use the fork to push food toward a sauce before eating it. Only small amounts of food should be taken at a time.

Some people like to emphasize their conversation by using their hands and arms. Servers should watch for this to prevent spillage and burns. Servers should indicate that service is coming so the guests may be warned. Guest should avoid pointing with utensils or waving them about, especially if there is food on the item.

When the menu is preset, such as at a banquet, tables will be set with all the flatware needed. At a formal meal, guests might sit down to as many as twelve pieces of flatware and wonder, "Which do I use first?" "What piece goes with what course of food?" Most people feel too embarrassed to ask. As a rule, the first utensil used is the one placed at the outer right or outer left and then the order of use is inward, so that just before the last (dessert) course is served, only the dessert spoon and fork are left (which can be at the top of the cover and not to the left or right). Sometimes, however, the first utensil to use is placed with the food item. Thus, a shrimp cocktail might be placed in front of a guest with the small cocktail fork on the plate. In glassware, one also follows the rule, outer to inner.

Sometimes guests are confused as to whether one should use the right or left outer item. Servers should help guests to know which to use. Although it is usual to place flatware in sequence of outer right or left in as courses proceed, there can be an exception, as in the case where the teaspoon may be placed on the outer right with the soup spoon inside next to it. The first course is a soup. One then does not use the teaspoon but picks up the soup spoon and uses that.

When a guest is finished eating, the knife and fork can be placed on one side of the plate, both pointing the same direction. A more traditional way is to leave the two utensils crossed in the middle of the plate. It is wrong to place used flatware on the table. All items should rest on plates. It is also wrong for servers to expect guests to reuse flatware. Fresh, clean items should be brought with each new course.

When drinking from a glass, the fingers should not be allowed to touch the rim. Goblets and stemmed glassware should be handled by the stem.

Etiquette for Specific Foods

Many foods require specific treatment to eat properly. Certain utensils, dishes, and other items may be required to proceed correctly along with special kinds of service. (See **Exhibit 11.1.**) A lobster bib, finger bowl, or special piece of flatware may be used. Some foods must be accompanied by special condiments or sauces. Servers should know the specific needs of these different foods so people can eat them properly. It may also be that by just serving things right servers can do much to aid guests in eating a food item properly, even if they're eating it for the first time.

If a guest does not know how to eat a food, it is not wrong to ask. However, if the guest prefers not to, he or she should wait and see what the others do, and then follow.

Finger Foods

Often at receptions, cocktail parties, and other social events, guests eat and drink standing up. Guests must sometimes have the skills of an acrobat to hold, eat, and drink. Party planners should have tables and chairs situated so guests can sit and eat. Another way is to have a number of small stands on which guests can set their beverages while eating. If no tables are available, there should be plenty of small bus stands on which guests can place their empty items. Guests do not like to walk too far to dispose of these items and, if tables are not supplied, plates and glasses will be found on top of armchairs, plant containers, bookshelves, and other places where they should not be.

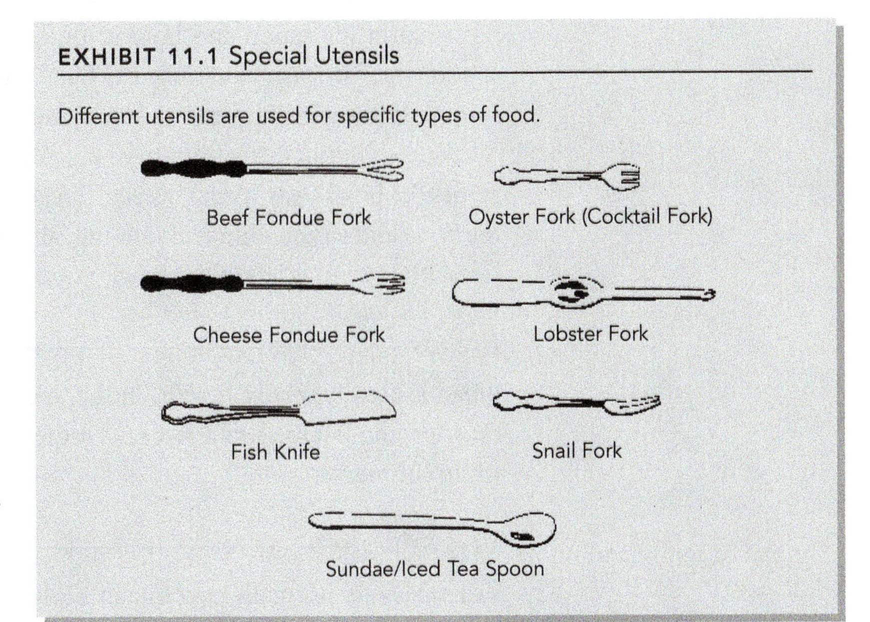

EXHIBIT 11.1 Special Utensils

Different utensils are used for specific types of food.

Moist or greasy foods soil guests' fingers, so plenty of napkins should be available. In some cases, servers might offer guests **finger bowls,** or small cloths dipped in hot water and then wrung dry.

Some foods are served very hot. If they are, servers should warn guests that they have just come fresh from the kitchen and to wait before biting into them. If such foods come with a small pick in them, they are easier to handle. Such picks also avoid soiling the fingers.

The plate most used at cocktail and receptions is the five-inch bread-and-butter plate, with or without a doily. Guests may pick these up and serve themselves from a buffet, or the food items will be passed by servers moving through the group. (Russians in the nineteenth century called the trays on which such foods were carried on *flying platters* because they were carried above guests' heads.)

Dips and sauces are often served at parties. After an item is dipped once, it should not be dipped again. Guests must take care to avoid spillage.

At picnics, foods such as hot dogs, hamburgers, submarines, dagwoods, and tacos may be served. It is often helpful to serve such foods wrapped in a napkin to hold them together and avoid spillage. The same is true of many desserts, such as frozen ice cream sandwiches.

Beef and pork ribs can and should be eaten with the fingers. When covered with sauce, ribs are very sticky, so extra napkins or a moist small towel should be provided. Other barbecued meats may also be considered finger foods and should receive similar service. A finger bowl should be served at the end of eating these items, if the operation uses them.

If shish kebob or other skewered item is served, it is proper to lift the skewer by its handle and push the pieces of food onto the plate with a fork. The skewer can be set aside on a plate.

Soups

Soup are served in a bowl or a cup. If the cup has two handles, the guest can lift the cup when only a small amount is left. This works only for thin soups. Drinking from a bowl is considered uncouth, as is blowing on the soup to cool it. Instead, take some hot soup in the spoon and hold it until it cools. Croutons should be served with a spoon when serving; never use the hands. It is better to eat crackers separately with soup, but it is not improper to break them up and scatter them over the soup.

Soup spoons are larger than other spoons. To use a soup spoon properly, move it across the bowl away from yourself. To get the last spoonful, it is not improper to tip the bowl and let the soup run into the far side where it can be spooned up. When finished, the spoon is left in the bowl, not on the table.

Breads

Bread is usually served as rolls, in slices, or in whole loaves. Loaves should be served on a cutting board with a serrated knife and a napkin with which to hold the loaf while cutting.

Vegetables

Vegetables are normally easy to handle, but a few can represent a challenge in eating. Corn on the cob is one. After adding seasonings, the cob can be held with one hand, both hands, or small cob holders stuck in the ends. Artichokes are eaten by pulling the leaves off one at a time, dipping the tender end into a sauce, and removing the soft flesh with the teeth. The leaf is then placed on a plate, which should be provided. The heart, or core, contains a layer of thistles over it. The diner can scrape these away and then eat the heart using a knife and fork. Asparagus can be picked up and eaten with the fingers, but it is more proper to use a fork and knife.

A vegetable may be served with its sauce on the plate, or the sauce may be brought to the table in a sauce boat. The server should ask guests if they want sauce and how much, or leave the boat so guests can add more if desired.

Salads

When eating salads, a knife may be used to cut large leaves, vegetables, and meats. Sometimes the salad is served as the first course. If guests prefer to eat it with their main meal, the server should move the salad to the guests' left. Some guests will eat some of the salad as the first course and then ask the server to leave it to be finished with the main course. The server should then move the salad to the left. If guests ask for salad dressing on the side, the server should bring it in a small container and place it on the side of the salad or on a small plate.

Pasta

Some people prefer to eat long pastas by twisting a small amount of strings around the fork and supporting it with a large spoon. Or the fork can be placed down almost vertically onto the plate and twisted until the desired amount is obtained. Others may choose to take a certain amount and bring it to the mouth without twisting. This might be more difficult considering that by rolling the pasta around the fork the size of the bite becomes smaller, more compact, and easier to eat.

Fish and Shellfish

Seafood can be tricky to eat properly. Some fish must be boned. Crustaceans must be removed from their shells. Often, guests avoid ordering these items, but servers can help them to overcome the difficulty and enjoy something that they really like.

Fish have a central bone structure to which most other bones are attached. The fins also carry bones. Often these are removed by filleting. However, if this is not done, these must be removed after service. Fin bones are easily removed by placing a knife or fork under the flesh surrounding the fin and removing it. By placing a fork or knife along the back and lifting, the flesh on top of the central bone structure can be exposed and this bone lifted out. In some small fish, small bones remain which must be watched for while eating. Often a fine-dining restaurant will have the chef de rang, captain, or maitre d' remove the bones, relieving the guest of this process.

In fine-dining operations, a fish fork and knife are provided. Although the knife can be distinguished easily from a dinner knife, a fish fork can easily be mistaken for a salad fork. Lemon wedges or halves are customarily served with fish or seafood. Special attention should be paid when squeezing lemon. If a wedge is served, the fork tines should be inserted in the center and the pressing should be light while holding the wedge between the index finger and thumb. It is recommended to place the hand as a cover so as to prevent the seeds and juice from squirting other guests. Some operations offer lemon halves wrapped in cheesecloth lined with a rubber band to facilitate neat squeezing. Drawn butter may be served with fish or seafood. It is proper to dip items in the butter, letting any drip fall onto the dinner plate.

Oysters and clams are often served raw on the half shell, often loosened from the shell for guests' convenience. They usually come on a plate set in a circle resting on a bed of ice with an oyster fork. A cocktail sauce will usually be in a small container in the center. Lemon or warm lemon butter or other sauce may be served with them. Some guests also like to have a hot sauce in addition. It is improper to cut raw clams or oysters. They should be eaten whole. One can dip them into the cocktail sauce, or take a bit of sauce with a fork and drop it on the clam or oyster. It is acceptable to crush oyster crackers into cocktail sauce.

Steamed clams and mussels are often served in a large bowl along with an oyster fork, sauce, warm lemon butter, oyster crackers, and some of the liquor obtained in steaming. One can pick up the shell and remove the meat with a fork or leave the shell in the bowl and remove it from there. The former method is preferred. It is not improper to lift the shell and remove the meat with the fingers, but this especially messy. The item is then dipped into the sauce or lemon butter. An empty bowl should be brought to the table for the empty shells. It is usual to follow steamed clams and mussels with a finger bowl.

The most difficult task in eating crustaceans such as lobster and crab is removing the meat. Shrimp and lobster tail are easier to handle, as their shell comes off little effort. To eat a whole or half lobster, hold the claw with a firm grip, preferably wrapping it in a napkin or holding it with tongs, or better yet, cracking it before serving. With the other hand, crack the shell in various places using a cracker, which should be provided. Insert a cocktail fork or pick and gently pull the meat out of the shell. It is not improper to remove the legs and gently suck on them. The roe or coral (eggs) of a female lobster is prized and can be eaten along with the meat. Warm drawn or lemon butter is usually served in a small dish or a plate. The server should place an extra plate or bowl at the right side of the guest for empty seafood shells. A lobster bib is appropriate not only for lobster but for any seafood item that must be shelled, but the server should give the guest the option of wearing one or not. A finger bowl with a thin slice of lemon should be provided after eating lobster.

If shrimp in a cocktail are large, there may be a problem eating them, since it is not correct to pick up the shrimp with a cocktail fork, bite off a bit and put the shrimp back. If they are hung around the outside edge of the cocktail glass, they may be removed with the cocktail fork and set onto the plate. They can then be cut with a fork and the pieces dipped into the sauce in the glass. If they are in the glass, it is proper for one to hold the glass and cut the large shrimp into bite-sized pieces. If sushi or ceviche is served containing large pieces of fish and seafood, it is proper to cut them into bite-sized pieces.

Desserts

In fine-dining establishments, the dessert fork and spoon are set together above the cover pointing in different directions. In many other establishments, these utensils are brought with the dessert. Some desserts can be eaten either with a fork or a spoon. The server should bring both utensils so the guest can use either.

Fruit is sometimes served as a dessert, usually sliced or cut somehow for easier eating. If this is done, guests will use a fork for some fruits such as a piece of melon and a sharp fruit knife for whole fruits such as apples, peaches, pears. A plate should be available for a diner to leave peels, cores, and so on. A small sharp knife should be on the plate. It is not wrong for guests to pick up a fruit like an apple and eat it holding it in the hand. This is called *eating hand fruit.*

At the end of a meal when one is ready to leave, one should catch the attention of the server and ask for the check.

TIPPING

The question of tipping is new for every meal. Tips should be based on the bills total *before* tax. A 10 percent tip is the least one should tip for adequate service. Below 10 percent indicates to the server displeasure with the service. Fifteen percent is the normal tip for satisfactory service, although some tip more than that. The tip many be higher than 20 percent for unusually good service. If servers are to split a tip, the tipper should indicate the division. Fifteen to 20 percent of the nontax total is often given the captain, but some just slip them a $10 or $20 bill. Remember that people from outside the United States are not used to tipping.

The usual tip for valet parking is one or two dollars, but in some instances it might be higher.

Maitre d's also can be tipped; in fine-dining establishments they will expect it, particularly if an extra service is provided. An attentive server should not have to be asked for the bill.

Dining Etiquette of Various Cultures

Rules of dining etiquette are not universal. Cultures have vastly different dining rules; from cultures that use no utensils to cultures that use many sets of utensils to accompany various courses. What's more, dining rules are not always consistent within a culture. Different food sources, living conditions, and differences among people of a country make for different dining practices.

Ethnic dining in this country has become an integral part of the foodservice industry. The most common cuisines are Italian, Chinese, and Mexican and more establishments are offering Indian, Japanese, Middle Eastern, and other ethnic foods. The following synopsis details some of the most important differences associated with the food and service of various ethnic cuisines.

Traditional Jewish Service

Jewish dietary practices are interwoven with their faith. Jewish religious law dictates food selection, preparation, and service. Jewish holidays and religious events are often by marked with the service of special dishes.

Kosher literally means fit or proper, and refers to a set of dietary rules called *Kashrut.* These laws are biblically based, and their application has been determined by rabbinic interpretation.

Mammals must have split hooves and chew their food. Only a specific list of birds are considered Kosher. Both birds and mammals must be slaughtered in a specific manner, called *shechita,* by a trained Jewish slaughterer called a *shochet.* Furthermore, the meat must be *Koshered,* or soaked and salted in a specific manner to remove all blood.

Only fish with both fins and scales are kosher. This excludes shellfish. Fish do not need to be killed in any specific manner.

Dairy products and meats cannot be cooked or served together. There are separate cooking utensils and serving dishes, for meat and dairy. During the Passover holiday, no dish, utensil, plate, or other item that has touched a leavened product can be used for food preparation or service. Some foods are called neutral (*parve*), such as fish, eggs, vegetables, and fruits, and can be served with either meat or dairy.

No cooking is permitted on the Sabbath, so food must be eaten cold or prepared ahead and kept warm. In many foodservice operations, boilers furnishing the heat in steam tables can be used to keep food warm.

A blessing is said before the meal begins. On the Sabbath and holidays, there is always a special blessing said over bread, usually *challah,* a braided egg-bread.

European Dining Etiquette

For the most part, dining etiquette in Europe is similar to the etiquette in the United States.

Traditionally, European dining was more formal than American dining, but even Europe is becoming more fast-paced and casual.

Chinese and Japanese Dining

Although there are many differences between Chinese and Japanese dining etiquette, both emerged thousands of years ago in common. Chopsticks are used in both Chinese and Japanese dining. Some common food items found in both cultures are rice, wheat, soy sauce, and tofu (bean curd). However, the separation of these cultures for so many years has produced many differences in Chinese and Japanese dining etiquette.

Chinese Dining

Chinese dining is an adaptation of family dining, where food is brought to the table in serving dishes and guests help themselves. Often the foods are placed on a circular movable platform set in the middle of the dining table. Guests rotate this to get to the various dishes. The main plate is usually about the size of a salad plate, and portions are small. There is also a rice and a soup bowl. A rice wine may be served in small cups. There is typically no water or other liquid on the table because soups supply the liquid. Tea is sometimes served. Chopsticks are the main eating utensil, and a special soup spoon is used for the soups. Foods are usually cooked in bite-sized pieces so they can be picked up with chopsticks. If the piece of food is larger than a bite, the plate may be raised, and the piece picked up with the chopsticks for a bite. It is not considered impolite to raise the rice bowl to the mouth to guide the rice in with the chopsticks.

Traditionally, the host sits with his back to the door, with the main or honored guest opposite him. This custom springs from an old legend, where an emperor was visiting a friend. The emperor had many enemies, and he was in constant fear of being murdered by them. To relax him during meals, the host sat with his back to the door and placed the emperor opposite him so he could see who entered. The strategy worked. Enemies did come in, but were seen as they entered. The emperor was always ready for them. Today, the tradition holds.

Next to the main honored guest may be other guests seated in turn around the table toward the host. It is considered an insult to the main guest to have the grain of the wood running toward him; it must run toward other guests.

Japanese Dining

The Japanese take great care in the presentation of their food. Meat, vegetables, fruits, and other items are cut and served attractively. The color and shape of the dish on which the food is served—one dish for each food item—must suit the food. Placement of the food on the dish is precise and decorative. Glass dishes are often used in the summer, green porcelain in the spring, heavy stoneware in the winter, and basket-shaped dishes tinted with the color of autumn leaves in the fall.

Traditional Japanese dining is quite formal. One is expected to use the best manners, giving deference to others at the table. The meal is usually served all at once. One should start a meal by taking a grain or two of rice and then a small sip of soup. It is considered bad taste to have any cutting utensil on the table, so all food should be served in bite-sized pieces. Soups and beverages are sipped from their dishes.

Traditionally, the Japanese do not sit at tables or chairs to eat. Instead, they kneel on a *tatami,* or mat, and the food is served on a small, raised stand in front of them. For a group, a large, low table may be used instead of the individual stands. The host faces guests, and no one is ever seated on the ends. Some Japanese restaurants provide tables that sit above a shallow opening in the floor so guests can sit on the floor and let their legs hang down in the opening.

Often a dining area will have a *tokonama,* or a scroll on the wall. In front of this sits a low table with a Japanese flower upon it. The honored guest is invited to sit in front of the tokonama. However, the guest should refuse this honor, and only accept the invitation upon the host's insistence.

Many restaurants provide small ceramic holders for chopsticks. In places where these are not used, the chopsticks will usually come wrapped in paper. One should use the paper wrappings to rest the chopsticks. Pickles are typically served at the end of the meal.

Indian Dining

Indians are largely vegetarians. Some use no eating utensils, but eat with their right hand. The food is usually ladled out of bowls onto plates. Many people sit on the floor to eat, and no one wears shoes in the dining room. Food services have places outside where those going to eat can wash their hands. The upper classes usually eat with knives, forks, and spoons. Rice is eaten usually instead of potatoes. They avoid pork and most (of Hindu faith) do not eat beef.

Middle Eastern Dining

In dining the peoples of Middle Eastern countries, most of whom are Muslim, follow traditional religious influences, food resources, and social customs. It is customary to wash the hands before and after eating. One eats only with the right hand, and often no eating utensils are used. As many as thirty different dishes may be served in small quantities, with cold foods coming first. Guests serve themselves. Pita bread is commonly used to help lift the food to the mouth. When meat is eaten, it is usually followed with rice. If a knife and fork are used, these should be left in a crossed position on the plate when finished.

Guests are honored by being seated to the right of the host. Traditionally, pillows are placed in a circle around a table, and guests sit cross-legged on a pillow on a rug while eating.

CHAPTER SUMMARY

The tradition of table etiquette in Western countries has evolved since the Middle Ages in Europe. Writers have collected and published established customs but the customs themselves have developed over time by common adoption. Etiquette differs greatly from culture to culture, and from occasion, time, and company. At its core, etiquette is based on being considerate of other people and ensuring pleasant social interactions.

In a seated-service, white tablecloth operation, dining etiquette plays a much more important role then in the casual drop-in operation. Preservice etiquette includes taking reservations, providing a coat check, and allowing for seating preference. Once the guests are seated, servers must ask guests for a drink order, and announce specials and their prices.

It is helpful when a server knows how one should eat a food and what proper manners are when eating. Servers must know how to help guests in various situations, including allergies, inedible foods, passing dishes, and when to eat. It is important for servers to know what is needed in service so people can enjoy eating in public without embarrassment.

Utensils are always given precise placement on the table. Up to twelve pieces of flatware may be set. For flatware and glassware, the general rule of outer to inner should be used. It also helps for the server to know which supplementary utensils are needed for each of the specific items. When a guest is finished eating, the knife and fork can be placed on one side of the plate, facing the same direction, or placed in the middle of the plate, the rim of the plate acting as a frame.

Some foods require special utensils, condiments, or sauces to be eaten correctly. At functions where there are no tables and quests stand to eat and drink small stands may be available for guests to place their beverages upon while eating. Napkins and small picks are often served with greasy foods, or foods that are hot, to make the foods easier to handle and to prevent soiling. If dips and sauces are served at parties, guests may dip the item only once.

Soups, vegetables, and seafood all require special methods or additional utensils when eating. Soups are served in either a bowl or a cup. The soup spoon is larger than other spoons and should be moved across the bowl away from the diner. Vegetables like corn on the cob can be held with both hands, or preferably with attached cob holders.

Seafood can be very tricky to eat properly. Some fish must be boned, while crustaceans must be removed from their shells. Lemon, drawn butter, and cocktail sauce are often served with fish and shellfish.

At the close of the meal, the appropriate tip should be left for the server, and should be based on the total bill before tax. A tip between 15 to 20 percent is traditional, and if necessary, should be split to suit the service. In high-scale operations, it is typical to tip the captain and possibly the maitre d'.

KEY TERMS

etiquette
finger bowls
Galateo
Russian technique
Traite de Civilite
A Treatise on Manners
zig-zag method

CHAPTER REVIEW

1. Why is it important for a server to know proper table manners?
2. How did the rules for proper eating develop? Do they change over time? Are they inviolate?
3. A guest orders broiled lobster. What should be served with it?
4. What is a finger bowl used for? What is its proper service? When should it be served?
5. How should guests place flatware on the plate when they are done eating?
6. There is a rule saying that one never places the elbow or elbows on the table. There is another rule saying that guests should not use the fingers to bring food to the mouth. Can either of these be violated? Under what circumstances?
7. A guest asks for a large tablespoon along with a fork to eat spaghetti. Is the guest wrong in asking? Why or why not?
8. How does a guest displeased with the service indicate this in the tip? What is a low tip percentage? What is a typical tip percentage?
9. What should a server do if a guest spills something?
10. After the guests are seated, what is the first question the server should ask them?
11. Answer the following guest questions:
    a. "What are these little finger bowls for?"
    b. "Is it okay if we share something?"
    c. "Is it rude to start before everyone at this banquet is served?"
    d. "Which fork do I use for the salad?"
    e. "How do I eat these snails?"
    f. "Where do I put the mussel shells after I eat the mussels?"
    g. "Can I use my fingers to eat these ribs?"
    h. "What's the best way to get the last of the soup out of this bowl?"
    i. "What's the best way to pick up spaghetti?"
    j. "What is the proper way to eat these raw oysters on the half shell?"

Case Studies

Serving Fish and Seafood

John is an experienced server and secures a job in an upscale fish and seafood restaurant, the top restaurant in Boston of its kind. Besides ordinary knives and forks, it serves with its fish and seafood items the correct utensils, dishes, and tableware. He comes to the restaurant one day ahead to see the service and learn the proper dishes and utensils that go with the various fish and seafood items. What must he learn about serving lobster, oysters or clams on-the-half-shell, snails, whole fish that the server must debone for the guest, and so on? Write up what he records about serving these items.

The Tipping Problem

Mary is a good waitress and makes good money in tips. She serves a group of people. They want individual checks. The bill of one person comes to $19.50 plus a 15 percent tax or for a total bill of $21.43. The person pays by credit card. When the time comes for this individual to add the tip and give Mary a signed sales slip, she adds $2.15, or about 10 percent of the total bill. As noted, Mary is a good server and gave this table excellent service. What's wrong in what the individual did?

GLOSSARY

A la carte menu	Menu on which items are priced individually
Aboyeur	Announcer who receives orders from servers and places them with the kitchen staff
Action station	Buffet service consisting of cooking or "finishing" some food items in medium-sized pans over portable réchauds as guests go through the buffet line
Alkali	Substance used to treat chocolate to give it a light appearance and smooth texture
Anejo/muy anejo	Label description for aged light rum
Apéritif	Fortified wine flavored with herbs and spices and usually served before a meal
Arabica	Type of bean produced by the coffee shrub
Armagnac	Brandy from the Armagnac region in France
Aroma	Fruity or flowering scent of a young wine
Black tea	There are three kinds: green, oolongand and black; green tea is not fermented before drying, oolongand is partially fermented, and black is fully fermented before drying
Blend	Mixing method in which ingredients are mixed in an electric blender
Blended whiskeys	Straight whiskeys blended with grain neutral or other spirits
Blood alcohol content (BAC)	Standard established to judge when intoxication from alcohol occurs
Blush (rosé)	Pink wine that is light and fresh with a fruity flavor
Bock beer	Type of beer with higher alcohol content than lagers and heavier in body, richer in flavor, darker, and sweeter
Bonded	Produced from a single distillation at 160 proof or less, bottled at 100 proof, and unblended
Bottom-fermented beers	Lagers that are fermented slowly from the bottom

Boulanger	Parisian who opened the first restorante on the Rue des Poulies
Bouquet	Fragrance imparted by the winemaking and aging process
Brand-name (proprietary name)	Wine belonging exclusively to a vineyard or shipper who produces and/or bottles it and takes responsibility for its quality
Brut	Driest champagne
Build	Mixing method in which each ingredient is added one at a time
Cacao tree	Semitropical tree from whose bean chocolate is derived
Cahveh	Name given to coffee by the Turks in the Middle Ages
California menu	Listing of snacks and breakfast, lunch, and dinner items, all on one menu
Call stock	Brands that guests ask for rather than regular well stock
Calvados	Apple brandy from Normandy
Captain	Supervisor of the chef de rang and commis de rang who may do special work, such as deboning a fish or preparing Crêpes Suzette
Marie-Antoine Carême	Early great chef who served royalty and notables.
Cash bank	Money server corner to make change when customers pay their bills.
Champagne	Best-known type of sparkling wine
Chef de rang	Main or front server in French service
Chef de trancheur	Carver of roasts, such as game, lamb, tenderloin, or poultry, in French service
Civil Rights Act	Law barring discrimination against employees because of race, color, religion, sex, or national origin
Clouding	Uneven color or texture of tea resulting from the breakdown of tannin
Coffea Arabica	Coffee tree, a tropical evergreen shrub native to Africa
Cognac	Brandy from the Cognac region in France
Commis debarasseur	Busperson who assists the chef de rang and the commis de rang in French service
Commis de rang	Assistant to the chef de rang in French service
Control states	States that handle the sale and distribution of liquor
Cordials (liqueurs)	Distilled spirits that have been treated with special flavoring ingredients

Crumber/crumb brush	Instrument used to removed crumbs from tables
Cycle menu	Menu that offers a group of meals for a specific period of times before changing to another group of meals. Yet another group of meals may replace the second group for a certain period of time, Eventually the first group of meals will again be offered, and the whole cycle is repeated.
Delegation	Empowering employees with responsibility and authority so that they can accomplish various tasks
Demi-sec	Sweetest champagne
Dietary Needs	The foods and liquids one needs to maintain good health
Dessert wine	Typically sweet, rich, and heavy wine commonly served after dinner
Dram shop laws	Laws that hold servers responsible to third parties injured or killed by intoxicated patrons
Du jour menu	Menu that is planned and written on a daily basis
Etiquette	In this text, manners used in eating
EEOC	Equal Employment Opportunity Commission
Empowerment	Management tool by which employees are given the power to make decisions that achieve higher standards of service
Auguste Escoffier	One of history's greatest chefs, who, along with César Ritz, operated fine hotels featuring fine dining and excellent service
Ethyl alcohol	Residue liquid that results when alcohol is formed
Expediter	Employee who acts as a communication link between servers and kitchen staff
Extra dry	Dry champagne
Extraction time	Length of time that water and coffee grinds are in contact
Fair Labor Standards Act	Law protecting workers from 40 to 70 years of age from discrimination actions, especially discharge because of age (and also protects teenagers)
Family and Medical Leave Act	Law requiring employers with 50 or more employees to offer up to twelve weeks of unpaid leave in any 12-month period for reasons related to family and personal health
Fermentation	The process of changing a carbohydrate into carbon dioxide and ethyl alcohol
Finger bowls	Small water bowls for rinsing the fingers at the table
Fire Management Functions	A listing of what to do in case of fire

Flambéing Pouring a flammable alcoholic beverage over food, warming it slightly and then igniting it.

Finish Impression left in the mouth after tasting wine

Flying platter Tray of beverages carried by servers to guests who are standing to be served, such as at a reception

Flying service Reception service in which beverages and hors d'oeuvres are presented on trays

Food covers Lids that keep foods hot and allow for plates of similar sizes to be stacked on top of one another

Fortified wines Wines with added alcohol, usually brandy

French Revolution The rebellion of the French people that introduced democracy in French politics

Galateo Book about good manners by Giovanni Della Casa, written in the form of an imaginary conversation between a wise old man and his nephew

General menu Main menu of a hospital from which special diets are planned

Generic Classification of wine by broad general type

Grand Reserve (vieux) Label indication meaning that most of the brandies in the blend are 20 to 40 years old and the youngest is more than 5½ years old

Gratuities Tips

Green tea Type of tea that is often treated or steamed to prevent full fermentation

Green zone Monitoring system classification describing a person with a low blood alcohol concentration

Greeter Person who welcomes guests

Guéridon Table on which preparation occurs in French sevrice

Guest check Forms on which guests' orders are written

Guest-check system Any system in which servers write out guests' orders in a legible and organized manner

Guide dog A dog trained to assist a blind person to move around

Guilds Associations formed by artisans and skilled tradesmen to help regulate the production and sale of their goods

Halal Islamic dietary laws governing food preparation

Heat lamp A lamp under which food is placed to keep it warm

Hops Beer ingredient that provides pleasant bitter flavor

Immigration Reform and Control Act Law making it illegal to hire aliens not authorized to work in the United States

Jigger Device used for portioning drinks

Job description (specification) List of tasks and responsibilities

Kaffa Region of Ethiopia from which the world *coffee* seems to have originated

Kosher Prepared in accordance to Jewish dietary laws

Lager method Method used to make pilsner beer, requiring a long period of cold fermentation

Le grand couvert (the great cover) The elegant and lavish style of French service where final preparation of dish is tableside

Lees (sediment) Matter that settles to the bottom of wine

Light-bodied rums Smooth spirits with a light body and sweet flavor, made from fermented sugar cane or molasses

Line organization Establishment in which responsibility and authority flow from the top down

Liqueurs (cordials) Distilled spirits that have been treated with special flavoring ingredients

Maitre d' hotel The person in charge of service in the dining room

Malt liquor Highest of the beers in alcohol content

Maltose Sugar formed from starch in barley

Catherine de Medici Member of one of Europe's richest and most powerful families who, as queen, started the growth of lavish and elegant French dining standards

Methyline chloride Solvent often used to absorb caffeine when making decaffeinated coffee

Mise en place Getting everything ready for the job to be done, and keeping things in good order as work is done

Mixology The art of mixing drinks

Nibs Small pieces of roasted bean from the cacao tree

On the rocks Term used to describe a drink served with ice

Oolong tea Type of tea that undergoes a process of fermentation before drying

Participative leadership Leadership style in which managers act as coaches to lead teams to success

Pilsner Light beer with a strong flavor of hops; most popular style of beer in the United States

Pivot system Standard system of guest order location

Porter Heavy, dark brown beer in which the sweetness of the malt remains

POS (point-of-sale) system Computer system that tracks guest orders, sales, guest counts, and other internal information

Preset keyboard Machine-operated system that requires only that the server touch a key to order an item

Privacy Act of 1974 Law which forbids employers from asking non-job-related questions which might be discriminatory

Proof Twice the percentage of alcohol in a product

Proprietary name (brand-name) Wine belonging exclusively to a vineyard or shipper who produces and/or bottles it and takes responsibility for its quality

Réchaud Small heater used to heat or cook food

Red zone Monitoring system classification describing a person who is intoxicated or in danger of being intoxicated

Restorante Type of establishment first introduced by the Parisian Boulanger, who claimed that the soups and breads he served were healthful and could restore people's energy

Reverse pyramid Version of the team effort approach in which managers are at the bottom of the pyramid and front-line employees are on top

Grimod de la Reynière Editor of the first gourmet magazine

César Ritz Hotelier, who, along with Auguste Escoffier, operated fine hotels featuring fine dining and excellent service

Robusta Type of bean produced by the coffee shrub

Role-play Training technique that allows servers to see and practice what is to be learned

Rosé (blush) Pink wine that is light and fresh with a fruity flavor

Jean de la Roque French botanical scholar and explorer

Russian technique Using a spoon and fork to lift and carry an item to a plate

Sanctuary Premises where guests are treated well by their hosts and every effort is made to see that guests come to no harm

Sec Sweet champagne

Sediment (lees) Matter that settles to the bottom of wine

Service controls Tools or techniques that lead an operation to achieving its goals

Serviette Small towel used by servers

Shake Mixing method in which ingredients are shaken with a hand shaker or mixed with a mechanical mixer

Shopping service Type of buffet service arranged so that only one kind of food is on each table

Show-and-tell To show something and discuss it's properties

Silencer Felt-padded or plush material placed under a tablecloth to quiet the noises of dishes and utensils and absorb spilled liquids

Sneeze guard Clear panels on the serving side of a table that protect food from contamination by guests passing by

Sommelier Wine steward

Sparkling wines Bubbling wines that usually contain 8 to 14 percent alcohol by volume as well as CO_2

Spindle method Order-placing method in which servers put orders on a spindle for cooks to remove

Spirits Beverages produced by the fermentation of grains, fruit, plants, and other products

Standards Specific rules, principles, or measures established to guide employees in performing their duties consistently

Still Nonbubbling

Stir Mixing method in which ingredients are stirred with ice in a mixing glass and then the drink is strained into a chilled serving glass

Stout Dark, full-bodied beer in which the sweetness of the malt remains

Straight Term used to describe beverage alcohol when the spirit is served alone

Straight whiskeys Whiskeys that have not been blended with grain-neutral spirits

Sweet table Elaborate buffet of desserts

Taberna vinaria Small eating and drinking establishment of the ancient Romans, from which we get the word *tavern*

Table d' hôte menu Menu on which items are priced together in a group, often as a complete meal

Taktacalah Quarter of old Constantinople (now Istanbul) where the first coffee houses were located

Tannin Strong, slightly bitter flavor of red wine

T'e Chinese word from which the word tea originates

Tea Aromatic beverage obtained by the infusion of tea leaves with boiling water

Thermopolium Stone counter with holes in which foods were kept warm, found in ancient Roman food services

Third-party liability Charge against a server and/or operation if a patron becomes intoxicated from drinks served in the operation and injures others because of the intoxication

Top-fermented beers Ales, stouts, and porters requiring only about three days to ferment

Touch screen Computer screen that is sensitive to human touch

Traffic sheet Form on which servers sign for order checks

Traite de Civilite Book by Antoine Le Courtin, which summarized all previous writing on table etiquette and offers the author's own advice according to the dining needs of the time

Tray jack Stand on which heavy trays are placed

A Treatise on Manners Notable book by Erasmus about table etiquette

Trichlorethylene Solvent often used to absorb caffeine when making decaffeinated coffee

Unionized operation Representative organization that acts for servers

Varietal Classification of wine by the predominant variety of grape used

Vieux (Grand Reserve) Label indication meaning that most of the brandies in the blend are 20 to 40 years old and the youngest is over 5 ½ years old

Voiture Small, mobile cart

Voyage de l'Arabie Heureuse Book written by Jean de la Roque which reports that the Turks made coffee popular in the Middle-Ages

VS/* ** Label indication meaning that brandies in the blend are less than 4 ½ years old

Wave system Banquet service system in which the entire room becomes one large station

Well stock House liquors used when no brand is specified

Yellow zone Monitoring system classification describing a person approaching intoxication

Zig-zag method Process used in the United States in which people cut their food and then transfer the fork from the left to the right hand

Duties of Service Workers

Duties of Greeters

Usually, a host or hostess is responsible for greeting guests, seating them, and often handing each a menu. This individual is also a supervisor of dining room employees: servers, buspersons, and others. Since this person is first to meet guests, the meeting must be cordial, friendly and give the guests a favorable impression of their coming experience of dining. The manner in which the host or hostess manages the dining room employees can have much to do with their job satisfaction, which then can filter down to the way service personnel handle guests, and to guest satisfaction with their dining experiences. One of their supervisory tasks is to see that dining room personnel present a proper professional appearance and duty performance. Other tasks are as follows:

1. Assign stations.
2. See that table appointments are correct and ready for guest arrival; check to see that the table and area around the table is clean, neat, orderly and inviting,
3. Help minimize dining room costs.
4. Check reservations and set up a plan for seating arrivals.
5. Balance workloads of servers by the way groups are seated.
6. Stand at the guests' right and give them their menus, handing with the left hand to the guests' right hand. If menus are not given out by the host or hostess, the server does it. The host or hostess sometimes names the specials available at that meal. While guests study the menu, a server, greeter, or busperson fills water glasses, serves butter, and so on. Others also gather menus after orders are taken.
7. See that disabled guests are given the help they need, children are given booster chairs, and babies have high chairs.

8. See that after seating, guests are met by their server and orders are taken and served properly.
9. Give any assistance to server needed.
10. Keep ashtrays clean and water glasses full.
11. See that items served are up to the establishment's standards, are at the proper temperature and are attractive.
12. It is proper for the host or hostess to pull out the chair to assist the guest in seating. In some operations, the host or hostess may pick up the napkins by the corner, letting them unfold, and then hand the napkins to the guests. This is done on a guest's right.

Duties of Servers

The server takes over from the greeter. The duties are as follows:

1. The approach to the table is one of smiles and good will. Greet guests, usually saying, "My name is _________ and I will be your server this morning (afternoon, or evening)." If the table is not preset, the server immediately brings the proper setting, seeing also that the center setup (e.g., salt and pepper) is in place. If the server is busy and cannot attend the table, some acknowledgment and apology such as, "I'm sorry I cannot help you now but I will be back as soon as I can to take your order," should be given.
2. Orders are usually written, except perhaps at a spot like the counter where the kitchen stands behind where the server is standing. Some operations use tel-service where servers can punch in orders into a small handset and send in their orders while standing in the dining area. Usually orders are taken clockwise with some fixed point as a start. Suggestive selling is in order. Servers should repeat the order as they write it down.
3. Iced water is next brought in and set at the top of the knife on the right of each guest. The glass is handled near the bottom.
4. If the table is not set, the server brings the necessary items when the water is brought and puts them in place with the water. (In many operations, a separate person pours the water.)
5. For breakfasts, the coffee is brought with the water. It is repoured frequently.
6. Servers should be prompt to suggest extra things such as a cocktail, soup or appetizer, dessert, and so on.
7. If guests seem to be in a hurry, suggest items that are ready.
8. Often servers write prices down as they write the orders.
9. After orders are taken, the menus are picked up and the orders are sent to the kitchen.

10. Serve cocktails and appetizers from the guest's left on a plate, doily, or napkin. Servers should suggest another cocktail when they see the first is nearly gone. The meal can be served when the guest has almost finished the drink or appetizer. Servers should know the correct items to serve with cocktails and appetizers. Thus, oysters on the half shell need an oyster fork, cocktails often need wafers or nuts, bouillon a soup spoon, melon a spoon, and so on.
11. As the order for a guest is finished, the server should thank the guest and move to the next guest
12. After the order is placed, turn the order in to the cashier to have a bill prepared. The order is then delivered to the kitchen for preparing.
13. An appetizer may start the meal, or sometimes a soup or salad with proper accompaniments or utensils such as a soup spoon and crackers for a soup, a cocktail fork with proper accompaniments for seafood cocktail, and so on. When these are consumed, the dishes and utensils are removed.
14. If bread or rolls and butter have not been served before this, they are when the main course is brought.
15. The main plate is served directly in front of the guest with the main item on the plate in the six o'clock position. The salad is served to the left, just above the fork. Condiments should be placed in the middle of the table where guests can reach them or have them passed by one guest to a guest that cannot reach them. Glance at the water glasses to see if they need refilling and fill them if they do.
16. A check back in a minute or two should be made to see that everything is satisfactory. Check rolls or breads, butter, and water. Do not lift glasses from the table when filling them.
17. Sometimes a table wants an individual's dishes and utensils removed when an individual is finished. Sometimes, the host or hostess will ask to have all removed when the last person finishes.
18. When the main course is removed, leave on the table the coffee cup, water glasses, cream and sugar, and coffee spoon. Ashtrays should be changed if filled.
19. The dessert orders are taken with each guest thanked as the server moves on.
20. The server sees that the check is written with tax and other charges written in as the last items. The bill is totaled. Some operations have a gratuity charge—usually 15 percent—added in. The check is on a small tray or plate face down. It is given to the host or hostess or placed in the center of the table if the person paying is not known. If payment is by credit card, this is picked up with the bill and taken to the cashier for verification. Both are once more brought back to the table, where the person paying puts down the gratuity on the bill. The server picks up the bill, thanking the guests

with a cheery, "Please come back again. We enjoyed having you," or something similar.

21. After guests leave, the table is cleared immediately and readied for the next guests, or set for the next meal if the table will not be used again that meal. The chairs should be brushed and inspected to see they are clean. Crumbs in wiping the table should be brushed onto a tray or flat plate and not on the floor.
22. A very important factor in being able to get a smooth flow of work is to have the service station well prepared and stocked before service begins. Extra dishes and glasses, condiments, ice, water, butter, and so on should be in place.

Duties of Buspersons

The word *bus* is taken from the word *omnibus,* which according to Webster's *New World Dictionary,* Desk Edition, means "providing for many things at once." This is an apt description of a busperson's job.

Proper grooming is important for buspersons as well as servers. Clothing should be neat, clean, and pressed. No excessive jewelry, gaudy makeup, tennis or open-toed shoes, or sandals should be worn. Some operations have uniforms that all people in serving positions are to wear, and the complete outfit should be worn properly. Shoes should be shined. Brown or black shoes are heavily preferred.

Servers often make most of their money in tips. Buspersons can do much to help them make good tips by helping them serve expeditiously and well, bringing food from the kitchen promptly for servers, filling water glasses, emptying ashtrays, and keeping fresh coffee brewing. They should keep tray jacks available and clean. Servers should reward their buspersons with part of their tips, often 10 to 15 percent.

A busperson's main job is to clear, clean, and set up tables, and this should be done quickly and well. Other tasks are refilling water glasses and coffee cups, helping to keep work stations filled with butter, lemon wedges, tea bags, adequate napkins, doilies, and place mats, and taking care of any soil or cleaning tasks. If you are working the breakfast shift and not the lunch, leave stations prepared for that meal and not for breakfast. Clean chairs, high and booster chairs, and dining area when guests have left. Crumb table with a brush or towel and brush crumbs onto a plate or tray, not on the floor. Stack trays properly, and do not overload them. Carry soiled items away immediately.

After removing items from the table and taking them to the kitchen, cleaning tables and area, resetting table after guests leave, buspersons should assist servers in any way necessary to do their work.

Duties of Cocktail Servers

The cocktail server should be on time, properly dressed, well groomed, and with the service station checked for extra trays, tip trays, matches, napkins, stirrers, straws, garnishes, ice, water, and so on. The first thing is to check the assigned service area to see if it is ready to receive guests. If a server is responsible for receiving payment for bills, the server should have the proper change and larger money ready. Several pens should be available.

The server uses a tray for all service to guests, carrying items with one hand except when carrying heavy loads. A bar towel is carried, draped over the arm.

Load trays in proper sequence according to the way guests must be served, except when heavy objects are to be carried. In this case, these objects are loaded in the center edge of the tray in the spot closest to the body.

Greet guests and seat them according to their desires, if possible. A cocktail napkin is immediately placed in front of each guest with the logo facing the guest. To avoid asking, "Who gets what," set up a system for identifying order to guest. Take orders in a clockwise fashion, sometimes ladies first, according to the policy of the establishment. Be sure to get the details right. There is a difference between bourbon with gingerale chaser and bourbon and gingerale. Record with the order details such "with a twist," "a splash of water," "on the rocks," and so on.

Bar setups differ, but usually there's a spot on the bartender's work area where servers stand and place their orders. Things like garnishes the server is to add to the various drinks are there, as well as extra straws, picks, stirrers, olives, cherries, and similar things. Often the server has to select the proper glasses and have them in place for the bartender. Chasers and mixes will often be there. Check glasses to make sure they are not chipped, are sparkling clean, and are free of lipstick. Some guests order a special brand such as a malt scotch, which is not a blended scotch. Be sure these differences are called distinctly to the bartender.

Bartenders are often busy with guests at the bar. They must sandwich your order between taking care of their guests. Wait for their acknowledgment and then give your order in an organized manner, such as, "scotch on the rocks; two drafts; one Miller light; one Chardonnay." The price of drinks is usually pinned on the wall at the order point; check to see that pricing is right. When paying for drinks, receive a check, which the bartender tears nearly halfway or more to prevent servers from presenting them twice or more.

Handle glasses by their base when serving, not near the top where the guest's lips might touch. Serve drinks directly in front of guests, chasers and mixers, etc on the right side of the drink. Serve from the right. Some guests want to pour their own mixes, beers, and other bottle goods, while others allow servers to do it. If necessary to reach in front of guests, excuse yourself. Check ashtrays.

Present bill face down on tip tray. Carry coin change and bills on a tip tray. Hold bill under change tray, when giving change. Put change on a tip tray in front of guest paying the bill. Call out the change transaction when putting change down, such as "Your bill was $23.50; $23.50 out of $30 is $6.50." This helps prevent arguments over change. Thank the guest paying bill. Thank guests when leaving, and cordially invite them back.

After that, clear the table and clean the area, straightening out chairs and picking up anything dropped. See that anything on the table is clean and ready for the next guests to use.

INDEX

C

T